OUT OF
THE IRON FURNACE

OUT OF THE IRON FURNACE

The Jewish Redemption from Ancient Egypt and the Delivery from Spiritual Bondage

by

ELIEZER BEN-DAVID

translated and adapted by
YAAKOV FEITMAN

SHENGOLD PUBLISHERS, INC.

New York City

ISBN—0-88400-040-0

Library of Congress Catalog Card Number: 74-84829

Published by Shengold Publishers, Inc.
45 W. 45th St., New York, N.Y. 10036

Printed in the United States of America

CONTENTS

TRANSLATOR'S PREFACE

In the winter of 1972, Rabbi Eliezer Ben-David delivered an extraordinary series of lectures in Yeshivat "Porat Yosef," Jerusalem. These lectures, which were soon published in Hebrew under the name *Mateh HaElokim* ("The Staff of the Lord"), evoked widespread admiration and praise from rabbis, scholars and critics.

The uniqueness of Rabbi Ben-David's book lies in his totally original approach to the story of the Exodus from Egypt. Although the subjects of the Ten Plagues, the Bondage in Egypt and related topics have long been explored in books on Jewish theology and homiletics, most previous studies have tended to be brief, aphorismic and usually limited to commentary upon the Passover Hagadah.

Rabbi Ben-David demonstrates, with great erudition and

eloquence, the interrelation of all aspects of the Jewish sojourn in Egypt: the rise to acceptance and respect, the descent into slavery and finally, the glorious redemption. In particular, Rabbi Ben-David sees in virtually every detail of the Ten Plagues, a different lesson for the future "nation of priests." Furthermore, each plague corresponds to the sins of a different period in history, all of which were encompassed within the Egyptian civilization. The plagues are therefore closely examined in individual chapters.

For a number of reasons, the entire book has been reorganized and adapted in the translation process. The original *Mateh HaElokim* was written hastily in a highly specialized Hebrew, which combines elements of the author's poetic phrasing and his vast erudition in both the Midrashic and Kabbalistic literature. Besides the direct quotations, there are countless echoes of phrases to be found in the Talmud, Midrash, Zohar and both Hassidic and Mussar works. The result is therefore a closed book to almost all except the very scholarly.

Yet, many who read and were impressed by the original, have felt that the thoughts and insights to be found in *Mateh HaElokim* should be made available to the general public. The title "Staff of the Lord" is one born of Rabbi Ben-David's modesty, for it only reflects a small part of Rabbi Ben-David's achievement in this work. We have therefore renamed the English version *Out of the Iron Furnace*, a phrase taken from the verse in Deuteronomy 4:20:

"But you has the Lord taken and brought forth out of the iron furnace, out of Egypt, to be unto Him a people of inheritance as you are this day."

The full significance of this title can only be understood upon arrival at the final chapter of the book, but for now let it suffice to say that although the *Furnace* refers specifically to the burning cauldron which was Egypt during the time of the Exodus, in the larger sense we all live in the *Furnace* of exile, in the conflagrant crucible of modern culture and mores, and often decadence and dissolution.

It is Rabbi Ben-David's gift to be able to show us, in graphic detail, that the seemingly strange combination we are witnessing today, of immense scientific achievement and deep cultural retrogression and immorality, is actually nothing new in human history. Ancient Egypt, also, was at the very height of its technological power, when it was at its lowest point in moral and ethical standards. The translator hopes that he has been able to convey some of the beauty and wisdom of the original.

A few points about the translation itself:

1). Because the original was prepared by Rabbi Ben-David from his lecture notes, many essential explanations and commentaries were omitted. The translator has therefore interpolated many entire sections not found in the original and takes full responsibility for their content.

2). In order to render the text as readable as possible, all references have been left to a chapter by chapter Bibliography to be found at the end of the book. Major citations, such as long quotations from the Bible, Talmud etc., have been left in the text proper. Scholars are invited to follow up the references in the Bibliography, where meticulous sources have been provided for almost all matters alluded to in the text.

3). Finally, the transliteration is into Sephardic Hebrew, as set forth in Volume I of the Jewish Encyclopedia. Translations from the Bible are generally from the Jewish Publication Society, 1955 edition, although all Middle English has been modernized. Quotations from the Talmud and Midrash Rabbah follow the Soncino but have been altered considerably.

Both author and translator wish to thank the publisher, Mr. Moshe Sheinbaum, for the patience, painstaking care and exceptional integrity he has shown with a difficult project. We wish also to thank Rabbi Moshe Greenes for reviewing parts of the manuscript and offering his valuable suggestions. Special appreciation goes to Rabbi David Cohen who gave generously of his vast fund of knowledge and expertise to the successful completion of this work. We wish also to thank Miss Etta Felman for contributing the index to this book.

It is the translator's fervent hope that this book will find its way into the homes and hearts of Jewish people everywhere and help all of us to emerge from The Iron Furnace.

Yaakov Feitman,
20 Sivan, 5735

INTRODUCTION

Y'TZIAT MITZRAYIM—THE EXODUS FROM EGYPT. Even as we utter the words, visions enter our minds of ancient glories and miraculous events. The collective memory of the Jewish nation crystalizes in our minds and we stand a bit taller. Yet, if we think of the Exodus only in terms of history, we miss the essence of that grand event. Our sages (Pesahim 116b) teach:

"IN EACH AND EVERY GENERATION A PERSON MUST REGARD HIMSELF AS HAVING PERSONALLY COME OUT OF EGYPT."

If we properly absorb the full meaning and significance of this teaching, it could change our entire philosophy of life. It should be obvious that this quotation does not mean that we, today, are free men and women because of the Exodus long ago. We make this thankful assertion much earlier in the Seder when we say, "Had God not taken our fathers out of Egypt, then we, our children and grandchildren would still be enslaved to Pharaoh in Egypt." Clearly, to regard ourselves as having *personally* come out of Egypt means much more.

In order to see the subject in its proper perspective, we should bear in mind the importance our sages attached to the daily "Remembrance of the Exodus." To cite but two examples known to every observant Jew:

1. TEFILLIN

Every weekday morning, all Jewish males must place the *tefillin* (phylacteries) on the head and arm. Before actually performing the commandment, the following meditation is recited:

"The *tefillin* contain four sections of the Torah which proclaim the absolute unity of God, blessed is His name, and remind us of the miracles and wonders which He did for us when He brought us out of Egypt." Thus, every weekday morning, as the Jew binds his heart, body and mind to the service of the Lord, he recalls the great events of the Exodus.

2. KIDDUSH

On the Sabbath, too, when we do not put on the *tefillin*, the Exodus is not forgotten. In the familiar, beloved prayer ushering in the Sabbath, *Y'tziat Mitzrayim* occupies a central position:

"The Sabbath is the first among the holy festivals which recall the Exodus from Egypt." As we chant the Kiddush sanctifying the holy day of rest, we keep firmly in mind that the Sabbath and all the festivals, commemorate the freedom gained at the Exodus from Egypt.

A further indication of the importance of the Exodus may be seen in the Talmudic (Berakhot 21a) ruling that "if one is not sure whether he has recited *Emet V'yatziv*, he must repeat the entire section, for it is a *D'oreita*—a Biblical injunction." Rashi explains that since this prayer mentions the Exodus, which by Biblical edict must be remembered daily, it is to be repeated in case of doubt.

Maimonides, too, stresses in a number of places the importance of remembering the Exodus. When discussing the laws of *K'riat Shma,* he notes that "although the *mitzvah* of *Tzitzit* is only performed during the day, we nevertheless read the portion in the Torah concerning *Tzitzit* at night also." He goes on to explain that this is done solely because the Biblical portion concerning *Tzitzit* also mentions the Exodus from Egypt, which must be remembered both during the day and night.

Recent Rabbinical authorities have shown that without even verbalizing the story of the Exodus, we can fulfill our Biblical obligation. As we perform the holy commandments so deeply connected with the Exodus from Egypt—*tzitzit, tefillin, mezuzah, matzah etc.*—our very bodies become living testimonials to the miracles and wonders of the generation when we left Egypt and became a nation.

Our great King David also teaches that the Exodus serves as a constant source of inspiration and a guiding light through the long darkness of exile. Thus, we sing in Psalm 111:4:

"He has made a memorial for his works; the Lord is gracious and full of compassion." Rashi, as usual, releases for us the inner beauty and profound significance of King David's words. He explains that the wording may be reversed in the following way: "Because He is full of compassion and wishes His children to be pious, He granted them the Sabbath and holy days, which remind them of their days of bondage in Egypt."

Rashi unlocks for us here a great secret about the holidays and *mitzvot* associated with the Exodus from Egypt. In the performance of these commandments, not only do we fulfill the Divine will, but we are automatically brought closer to true holiness and piety. Surely, every *mitzvah* we perform brings us closer to the Almighty! Surely every time we fulfill the will of God we become a little more pious! What is the special power of the *mitzvot* associated with the Exodus which causes them to make "His children to be pious?"

Rashi gives us the answer in the way he reverses the words in the passage from Psalms:

"Because He is full of compassion and wishes His children to be pious, He granted them the Sabbath and holy days, which remind them of their days of bondage in Egypt." The message is a clear and unmistakable one; every detail, every phenomenon, every manifestation of the Exodus is part of one vast lesson plan for humanity. No aspect of *Y'tziat Mitzrayim* is merely a footnote in history, to be remembered perhaps by an obscure scholar in a musty library. The Exodus is a living, throbbing *guide to life in the present day.*

In the pages and chapters that follow, we hope to illuminate those aspects of the Exodus which are particularly meaningful to our own generation. As we delve ever deeper into the sociology and culture of ancient Egypt, we find an uncanny similarity to those of our own generation. As we look back upon ancient Egypt, we feel as if we are gazing through a distorting mirror. The shapes are different, the sights are out of proportion, but the essential patterns are the same. As the drama of bondage and redemption unfolds before us, we shall discover the reason for this extraordinary similarity.

For the moment, let us utilize our imaginations to take a trip through time and space to an era in our history so laden with suffering and so rich in glory, triumph and ultimate redemption, the AGE OF THE EXODUS. The trip is difficult and the journey is a taxing one, but let us remember that in making it we fulfill the *mitzvah* of regarding ourselves as having "personally come out of Egypt." The journey will also reward us with a clearly marked map for travel through this difficult era.

Chapter I

THE BEGINNINGS OF BONDAGE

THE CHARACTER OF ANCIENT EGYPT

As we embark upon a comparison of our own age with that of the Exodus, we immediately encounter a crucial question:

"What can *we* possibly learn from the condition of the Jews in ancient Egypt? They were a people in bondage, miserably enslaved by a barbaric nation, while most of us are fortunate enough to be free citizens of civilized democracies, enjoying the fruits of liberty and the beneficence of advanced cultures."

The question is a valid one, yet the very premise that our ages are so far apart in character stems from a widespread misconception of ancient Egyptian civilization and the beginnings of the Israelite servitude.

First and most important, let us examine exactly what kind of a country ancient Egypt really was. Archeologists have unearthed a vast amount of detail on the "Rammeside Age" in Egyptian history, which coincides with the Exodus of the Jewish people from Egypt. All of the evidence points to a highly sophisticated well-developed civilization:

"The Egyptians . . . developed their own distinctive culture, which far excelled anything that man had achieved anywhere else," writes the noted Egyptologist Cyrus H. Gordon. In exploring the home libraries of ancient Egypt, J. M. Plumley discovered that "didactic treaties containing wise maxims and proverbial truths" were very greatly to the Egyptian taste.

Actually, we do not need archeologists to tell us of the highly "civilized" character of ancient Egypt. If we read carefully the beginning of the Book of Exodus, we cannot help but note that even after Pharaoh had decided that the Israelites would have to be enslaved or destroyed, he could not simply announce such a decree publicly. Instead, we find that he had to "deal wisely with them" (Exodus 1:10) and "with wicked intent" (Nehemiah 9:10). It is clear that the reason that Pharaoh had to "deal craftily" and in secret before he could openly enslave the Jews was that Egypt was too much of a democracy and considered herself too civilized for her ruler to suddenly decree slavery or death on an entire innocent people. How, then, did it all begin?

We can best appreciate the answer with a Halakha from the *Shulhan Arukh*, the Code of Jewish Laws. Although there are five varieties of vegetables which may be used to fulfill the *mitzvah* of *maror*—bitter herbs—on Passover, the most desirable is Romaine lettuce. The reason for this preference is that this type of lettuce "symbolizes very well the Bondage in Egypt. In the beginning it is sweet and then it turns bitter. So, too, were the Egyptians, soft in the beginning and hard at the end."

The Talmud (Sotah 11b) deduces this illuminating fact from the verse, "And the Egyptians made the children of Israel serve *b'farekh*—"with rigor" (Exodus 1:12). Our sages point out that the word *b'farekh* can be separated into the two words *b'feh rakh*—"with a soft mouth." How did this "soft mouth" manifest itself? In what way were the children of Israel lured into bondage?

PHARAOH'S TRAP

The Midrash enlightens us:

"One day, Pharaoh gathered together all the children of Israel and entreated them, 'won't you please do me a favor today?' Thereupon, Pharaoh himself picked up a basket and shovel and began the work of making bricks. Whoever saw Pharaoh doing this work went to join him. Immediately, the Children of Israel went with enthusiasm and joined him in the labor. Since they were naturally strong and able, they produced an enormous amount of bricks. At the end of the day, Pharaoh appointed taskmasters over them and ordered them to count the bricks. When they had finished counting, he ordered, 'this is the amount that you shall deliver to me every day.' "

Rabbi Haim ben Atar, in his holy commentary *Ohr Hahaim*, goes even deeper into the psychological forces that helped to shape the beginnings of the great Bondage. It is self-evident, he explains, that even before the Jews arrived in Egypt, Pharaoh must have had many workmen, builders etc. Egypt was already at the time the center of a thriving empire, where grandiose architecture was a prime preoccupation of the Pharaohs. With this thought in mind, the *Ohr Hahaim* reinterprets the passage in Exodus 1:11:

"Therefore did they set over them taskmasters to afflict

them with their burdens." The general understanding of this verse is that Pharaoh set *Egyptian* taskmasters over the Jews to ensure the completion of their work. The *Ohr Hahaim* points out that this method of enslaving them could certainly not be called "let us deal wisely with them," as the Torah records Pharaoh thought of it. Surely the assigning of cruel taskmasters upon an innocent people could only be thought of as barbaric and uncivilized.

Therefore, the *Ohr Hahaim* concludes that the verse means that Pharaoh appointed the *Jews* as taskmasters over all peoples who had previously worked for Pharaoh. This revelation sheds an entirely new light on the genesis of the Bondage in Egypt. Manifestly, when the Jews first began to perform the work for Pharaoh, it was considered by many a great and unexpected "honor." At first it seemed as if the Jews alone, from among all the various ethnic groups in Egypt had been chosen to *personally* work *with* Pharaoh on projects essential to the prestige of the Egyptian Empire. It seemed as if only the Jews were to be *trusted* with such sacred work and therein lay Pharaoh's cunning and "wise dealing." As the personification of Egypt's great fund of wisdom, Pharaoh discovered two factors within the psyche of the young Jewish nation which he exploited to the ultimate point of reducing the Jewish people to slavery:

THE TWO FATAL FLAWS

1). *PATRIOTISM*—Our sages teach that every impulse within human nature, every proclivity to which mankind is heir, every personal tendency or inclination, can be a positive or negative factor, depending upon how it is used and if it remains under rational control. Thus, although anger is usually considered an adverse trait, in the case of Phinehas, whose wrath and jealousy were for the Lord, his anger rescued the children of Israel from a destructive plague:

"Phinehas, the son of Eleazar, the son of Aaron the priest, has turned My wrath away from the Children of Israel, in that he was very jealous for My sake among them, so that I consumed not the Children of Israel in My jealousy." (Numbers 25:11).

At the other extreme, we know that King Saul was destroyed by the exercise of mercy at the wrong time, and the Talmud (Gittin 56a) even blames the destruction of the Holy Temple upon the misplaced modesty of a certain rabbi. Patriotism, too, is one of the traits which can be tremendously meritorious or totally destructive. Although our sages often praised the virtues of patriotism, they recognized that if taken to unhealthy extremes, it can bring disaster in its wake. So it was with Pharaoh's request for help from the Jews. If we reexamine carefully the Midrash quoted above, we must notice the special alacrity with which the Jews jumped to join Pharaoh in his work. The Midrash specificially divides those who joined Pharaoh into two groups: "Whoever saw Pharaoh doing this work went to join him. *Immediately*, the Children of Israel went *with enthusiasm* and joined him in his labor." No Egyptian subject could afford to bypass the Pharaoh working with his own hands and not contribute to the labor. But only the Children of Israel did so with *enthusiasm.*

2). PRIDE—When Pharaoh appealed to their vanity and offered the Jews the opportunity to sudden elevation as overseers and taskmasters, they readily assented. Although the Jews had never had any experience in building and many of them surely must have sensed the elements of artifice and chicanery, nevertheless, their sense of pride did not allow them to think rationally and they were victimized and out-maneuvered.

Clearly, the Midrash and the *Ohr Hahaim* wish to teach us that if it were not for these two factors, Pharaoh would have had a much more difficult time suddenly imposing slavery upon an innocent people. With what we can only term their own tragic help, Pharaoh was able to *ease* the Jewish people into a posture where they were known by all to be identified with and inexorable from the art and process of building. It was only an inevitable step from there to slavery.

THE CRUCIAL QUESTION

We must now ask ourselves that crucial and painful question which goes to the heart of *Galut Mitzrayim*:

How and why did the Jews, the noble descendants of Abraham, Isaac and Jacob, Sarah, Rebekah, Rachel and Leah, succumb so easily to the materialistic lures and wiles of Pharaoh? The solution seems even more distant when we recall that at the time when Pharaoh began to tempt them, the Jews were possessed of a special status in the land of Egypt. They lived in the fertile area of Goshen and were known to be kin to the famous Egyptian vice-regent and his venerated father. Indeed, Jacob's prestige in Egypt at the time of his death was so great that he feared that if buried in Egypt, his body would be worshiped as an idol. Why, then, did the Children of Israel jump to curry favor with Pharaoh? What did they think they had to gain and what brought them to this sad state of affairs?

THE FORTY-NINE LEVELS

We find the answer in the seminal words of the Holy Zohar:

"*I am the Lord your God who brought you out of the land of Egypt, out of the house of bondage* (Exodus 20:2).

Upon this verse, Rabbi Yessa, the junior member of the rabbinical college, said to his teacher, Rabbi Shimon ben Yohai: 'I have a question to ask of you, yet my heart trembles at the very thought. I am in a quandary; I am afraid that if I ask, I shall be punished, yet, if I do not, I shall forever be depressed with wonder and speculation.' Said Rabbi Shimon, 'ask.'

Said Rabbi Yessa: 'The Lord constantly reminds the Children of Israel that *I am the Lord your God who brought you out of the land of Egypt.* Why should the delivery from Egypt be considered such an extraordinary act of favor and grace? Surely the Lord had made a fast covenant with Abraham that *your seed shall be a stranger in a land that is not theirs . . . and afterward shall they come out with great substance* (Genesis 15:13-14). Why, therefore, is the Exodus constantly mentioned to Israel, as if to say, God granted us more than He promised to Abraham in the covenant?'

Answered Rabbi Shimon: 'come and see, my son, that the Holy One, blessed is He, only promised Abraham that He would deliver Israel from the bondage of Egypt, but no covenant was made to deliver them from the depravities of Egyptian idolatry. For this is certain, that when the Jews were in Egypt, they were contaminated and sullied by many sinister forces of evil and iniquity. Their moral condition steadily deteriorated until they had sunk to the forty-ninth level of impurity, the nadir of human degradation. The Holy One, blessed is He, not only raised them out of the depths of their uncleanliness, but elevated and purified the Jews until they reached the forty-ninth level of understanding and purity. This wonderful process of purification and spiritual elevation was totally unmentioned in the covenant with Abraham and was granted to the Jews solely through the grace of the Almighty.

For this reason, you will find the Exodus from Egypt mentioned fifty times in the Torah, to show the world the great favor that the Almighty did with Israel, of extricating them from the deepest defilement and raising them to the zenith of holy purity. In commemoration of that prodigious spiritual journey, we count the days and weeks between Passover and Shavuot, for on each day we emerged from one level of impurity and climbed a new rung on the ladder of understanding.' "

DO NOT FEAR TO ASK

We can well understand Rabbi Yessa's trepidation in approaching the subject of the meaning of the Exodus. Its ubiquity in the Torah and our liturgy makes it clear that the Bondage, Redemption and their aftermath are events that go to the very fiber of our belief and to the essence of our nationhood. For this reason, Rabbi Yessa was understandably reluctant to enter into the labyrinthine intricacies of the Egyptian exile. Yet, from Rabbi Shimon's readiness to discuss a difficult subject, we may glean a great practical truth about the ways of our Torah. It is always better to ask and to learn than to remain in ignorance, continue to wonder and perhaps come to the wrong conclusions. Yet, the great caveat is that we must be careful to whom we put the important questions of religion and belief. It is only the Rabbi Shimon ben Yohais of each generation who can guide us to a veracious understanding of the profundities of our holy Torah.

Rabbi Shimon places us on the proper path toward an understanding of the situation of our forefathers in Egypt. Clearly, the quintessence of the Bondage lay, not in the physical enslavement we usually associate with subjugation, but in the *moral* thralldom to which the Jews became victim. As we probe deeper and more perceptively into the condition of our ancestors in Egypt, we shall have to alter our conception of the meaning of exile and how it works its pernicious influence upon the human soul.

THE JEWISH SUCCESS STORY

The first misconception we must get rid of is the myth that the Jews were thrust in the early part of the exile into physical poverty and misery. The fact is that the majority of Jews lived in absolute opulence and luxury. As the Holy Zohar teaches us, the Jews in Egypt before the Great Redemption were steeped in the worst manner of evil and immorality. The precipice for their long slide down the forty-nine levels of defilement was their material success. We know that the sliding pond to iniquity begins only at great heights. The security of material wealth, the false tranquility of worldly success, lured the Jews then, as it has sadly so often throughout the generations, into a deceptive sense of belonging and kinship. As in the Greek exile many Jews became Hellenized, so had many Jews in the first *Galut* become Egyptized.

The Midrash reveals to us the extent to which the Jews had become successful in their foreign surroundings:

"It is written, *Vayagar Moav*—"and Moab was sore afraid." Another exposition of the expression *vayagar* is that it bears the sense of stranger (*ger*). The Moabites felt themselves becoming strangers in the world. They thought: The Israelites went down to Egypt to sojourn there and took possession of it, letting houses to the Egyptians; as it says, 'Every woman shall ask of her neighbor, and *migarat beita*—of her that sojourns in her house' (Exodus 3:22)."

We may deduce from the words of the Midrash that in antiquity, it was a universally acknowledged fact that the Jews in Egypt had achieved phenomenal success in their business endeavors, so much so, that in general, the Jews were the landlords and the Egyptians the tenants. This business success and all-consuming entrepreneurship later caused the Moabites to be apprehensive even at the approach of the peaceful Israelite camp.

Another Midrash is even more specific in depicting for us the character and aftermath of the Jews who learned the Egyptian life style and copied their decadent ways:

"Why did God, blessed is His name, before whom there is no respecting of persons and who searches the heart and knows the secrets of men, bring darkness upon them? Because there were transgressors in Israel who had Egyptian *patrons and lived in influence and honor and were unwilling to leave. . . .* On this account did He bring darkness upon the Egyptians for three days, so that the Israelites should bury their dead without their enemies seeing them, and for this they should praise God."

What percentage of the Jews lived in affluence? Was it only a minority which had fallen into Pharaoh's trap? Again, the Midrash is our guide:

"*Vahamushim alu b'nai Yisrael*—'And the children of Israel went up armed out of the land of Egypt' (Exodus 13:18). This word (*vahamushim*, from the word *hamesh*, meaning five), comes to teach us that only one out of every five Jews left Egypt. Others say, one out of fifty, and still others say, one out of five hundred. Rabbi Nehorai exclaimed; 'not even one out of five hundred were in the salvation, for it is written: 'I cause you to increase (*revavah*—ten thousand), even as the growth of the field' (Ezekiel 16:7). Furthermore, the Torah tells us that 'the Children of Israel were fruitful and increased abundantly, and multiplied, and waxed exceedingly mighty; and the land was filled with them' (Exodus 1:7). This great population increase was due to the fact that the Jewish women gave birth consistently to sextuplets. It is therefore impossible for any more than one in five hundred to have left Egypt with the Exodus. The rest died during the three days of darkness, which the Lord provided expressly for this purpose.' "

Rabbi Nehorai's teaching radically alters our entire image of the conditions in ancient Egypt when the Jews were in bondage. Manifestly, the great majority of Jews were so well off in Egypt and so enamoured of Egyptian culture that they did not *want* to leave even after the miracles that were witnessed by the entire nation. How can we reconcile this revelation with the verses in the Torah telling of the "lives bitter with hard service" (Exodus 1:14) and the cries and sighs of the Israelites

(Exodus 2:23)? For the answer, we must examine the concept of slavery in all its various forms and determine which type of servitude is closest to that which the Jews suffered in Egypt.

THE THREE FORMS OF SLAVERY

1). THE BORN SLAVE—In a way, one who was born into slavery, who has lived all his life with and among slaves, and has never experienced any other kind of existence, suffers the least amount of psychological damage because of his servitude. He feels no need to examine and evaluate every order he receives from his master, for he has been taught from birth that it is his essence to obey. He cannot have the deep yearning for freedom of the once-free, for he does not really know what freedom is.

2). THE FREE MAN SOLD INTO SLAVERY—The person who has once tasted freedom and has been sold into slavery reflects an entirely different personality. His entire being is shaped by his longing for freedom. He carries out his master's commands, but only because he must, so that he is constantly doing only what is demanded of him. Any dedication or enthusiasm, any hidden reserves of mental and physical prowess, any talent or ingenuity he may have are reserved for the eventual try for freedom. He does not become deeply or emotionally attached to his surroundings in bondage and he makes no great attempts to excell in the performance of his dreaded tasks. He lives a consciously ephemeral existence, waiting for the moment when he can begin a free and independent life. He is coiled like a spring, waiting for the propitious moment to leap, yet he is serene in his slavery, knowing that it does not define his essence and that he will soon be rid of its yoke.

3). THE UNKNOWING SLAVE—Worst of all, and most tragic for the human spirit, is one who has worked and *slaved* for many years, thinking he was laboring as a free man for himself and his loved ones, only to discover suddenly and humiliatingly that he has been toiling for someone else. This person has done not only the bare minimum, but the physical maximum. He has dedicated all his energies, applied all his courage, tenacity and experience to the aquisition of material things.

And he has succeeded. He has ammassed an abundance of the cherished symbols of success and can afford finally to enjoy the fruits of his labors. Then, suddenly, the bubble bursts and the hitherto unknown master arrives with the bill of ownership. Not only does everything he owns belong to another, but he himself is an object of helotry. The abrupt realization that one's entire life has been wasted, that one's superhuman efforts must go to another, is one of the supreme psychological torments that a human being can experience.

THE BETTER OF TWO EVILS

The Jewish People have gone through many forms of slavery. There have been many periods when we lived under the aegis of cruel and inhuman tyrants. We suffered through blood libels, pogroms and concentration camps. We were persecuted and discriminated against, oppressed and tyrannized. But, because we knew our communities were at best temporary, our plans tentative, our very lives characterized by impermanence, we were able to deal realistically with the situation. Our holy deep-rooted instincts took over and we concentrated on the spiritual: the study of Torah, the education of our children, and progressing spiritually and morally.

Not so has been our history during the more "congenial" periods of our exile. As the years came and went and we began to feel ever more at home, as time went by and no one came up to us with a document of expulsion or a decree of inquisition, we drove our stakes deeper and deeper into the corporeal ground.

We became bankers and lawyers, doctors and professors, accountants and businessmen. We erected edifices to our success and monuments to our permanence. The more we felt *at home*, the more we began to look, sound and feel like the natives of the country of our exile. As we slowly shed the outer and inner trappings of our religion, the danger of total assimilation loomed ever closer. It was always at that point that God *rescued* us by having our oppressor remind us that we *are* different.

THE SPIRITUAL BONDAGE

And so it was in Egypt. The Jews were sinking lower and lower into the unhealthy atmosphere of Egyptian civilization. Their financial security in the upper echelons of Egyptian society caused many to adopt the concupiscent Egyptian life style as well. Every day, hour and minute brought the Jews closer to the fatal forty-ninth level of impurity. Thus, the major danger to the Jews in Egypt, the essence of their Bondage, was not in their physical chains or the difficulty of their labors, but the mortal danger of their soul upon their near-assimilation into Egyptian culture.

Here we have the beginning of an answer to our difficulty concerning the character of the Bondage in Egypt. The terms *bondage*, *suffering*, *cries* etc. refer basically to the *Spiritual Bondage* in which the Jews were trapped. The Torah is not a history book of physical travail and triumph. It is an eternal guide to how to live the spiritually holy life. Therefore, the Torah does not relate details of physical suffering or pleasure unless they come to teach a lesson in morals or ethics. The meticulous detail with which the Torah records all aspects of the Bondage and Exodus indicates the great moral lessons they were meant to convey.

The Bondage in Egypt clearly corresponds to our third

category of slavery, the unknowing slave. Living luxuriously in the upper strata of the Egyptian social network, the Jewish people dedicated their energies to bettering themselves in their new society. They grew deeply entrenched in the mores and culture of Egypt and labored under thoughts of acceptance and permanence. At that point, the Lord shortened the prescribed period of their exile and prepared to deliver them from both physical and spiritual bondage.

Yet, in order to alert the Jews to the urgency of their moral condition, the Lord illustrated for them the folly of their infatuation with Egyptian culture. This was done first through a series of blows to their sense of acceptance into Egyptian society. Once they were isolated by the Egyptians *themselves*, they would be forced to think of themselves as an individual nation with unique laws and traditions. When this had been accomplished by adding physical to moral servitude, the fatuity of Egyptian culture was displayed to all through the Ten Plagues.

THE FAIRNESS OF THE TORAH

With this in mind, we are in a position to appreciate the extraordinary fairness of the Torah toward even our worst enemies. It would be so very easy for us to look back through our history and blame the Egyptians for having lured us into the trap of iniquity. It would be so tempting to claim that the ancient Egyptians used powerful persuation or even coercion to force the Jews to assimilate and follow in their evil ways. Yet, we find that precisely the opposite is the case. The Midrash teaches:

"Rabbi Shimon says: How do we know that one who causes a man to sin is even worse than one who kills him? Because one who kills him does so only as regards this world but leaves him a share in the World to Come. One who causes him to sin, however, kills him in this world and the next. Two nations advanced against Israel with the sword and two with temptation to transgression. The Egyptians and the Edomoites advanced against them with the sword, as is proved by the texts, 'The enemy said: I will pursue, I will overtake . . . I will draw my sword' (Exodus 15:9). Edom, too, similarly said, 'You shall not pass through me, lest I come out with the sword against you' (Numbers 20:18). Two advanced against them with the temptation to transgression, namely the Moabites and the Ammonites. Of those who had advanced against them with the sword, it is written, 'You shall not abhor an Edomite' (Deuteronomy 23:8); 'You shall not abhor an Egyptian' (Ibid.). Of those who advanced against them with the temptation to transgression, it says, 'An Ammonite or a Moabite shall not enter into the assembly of the Lord; Even to the tenth generation shall none of them enter . . . for ever."

At first, this Midrash seems to contradict all we have previously asserted concerning the moral depravity of Egypt during the Israelite servitude. If the Egyptian culture was indeed so decadent, why is Egypt classed with those who "advanced against us with the sword," and not with those who "advanced against us with temptation?" The answer becomes all the more important because we base our treatment of an entire people upon this characterization. The Torah identifies for us the essence of each nation's relationship with the Jewish people. If Egypt is regarded with the exhortation to the Jewish people, "You shall not abhor an Egyptian," then clearly the Torah does not blame the Egyptians for the fact that our forefathers emulated their evil ways. Why, indeed are the Egyptians not held accountable, as are the Ammonites and the Moabites, for willfully enticing our ancestors to transgression?

THE LURE OF EGYPT

Rather than a contradiction, the above-quoted Midrash provides resounding proof for our portraiture of the beginnings of the Bondage. The Egyptians did not have to *force* the Jews to adopt Egyptian culture, nor did they even have to make a formal attempt to influence them. The power of Egyptian culture was so all-pervasive, the lure of Egyptian immorality was so enticing, that just the *fact* of Egyptian culture was enough to snare the unsuspecting. The Ammonites and the Moabites had to mount intensive campaigns to tempt their would-be victims. All the Egyptians had to do was to be themselves. But, of course, for this the Torah does not blame them, and in fact commands us: "You shall not abhor an Egyptian."

THE ROAD TO ASSIMILATION

In the beginning, even those who succumbed to the lure of Egyptian culture thought they could also retain their Jewish identity. Until the very end, many of them retained some of their traditional Jewish clothing, ate only Kosher food and continued to speak Hebrew. Yet, it was not enough. Seeking to live in both worlds, the traditional Jewish one and the new glittering Egyptian one, many fell by the wayside. Even those who meant well and wished only to provide security for themselves and their families were eventually trapped by the powerful force of Egyptian decadence.

EGYPT IN THE MESSIANIC ERA

The fact that we do not blame Egypt for our moral decline during the Servitude has consequences that extend even to the Messianic era:

"Egypt is destined to bring a gift to the Messiah. He will

consider rejecting it, but the Lord shall say to him, 'accept it from them for they provided my children with a dwelling place.' " (Pesahim 118b).

Although the Talmud (Berakhot 63b) recognizes very well that anything the Egyptians did for the Jews, even when Jacob was alive, was purely for selfish reasons, we still do not totally reject the Egyptian nation, for we were in fact their "guests" for an extended period of time. These words of our sages confirm our assertion that Egypt did not willfully lure us into sin. Amon and Moab have no place bringing presents to the Messiah. Their temptations were willful and skillfully planned and they are eternally placed in ignominy. For our actions in Egypt, we can and do blame only ourselves.

THE ARCHETYPAL EXILE

We must now probe even deeper into the origins of the Egyptian Exile and ask ourselves why it should have happened the way it did. It is well to remember that in Jewish history there are no coincidences and no minor happenings. Everything is conducted according to a vast Divine plan and every detail must contribute to the symmetry and understanding of the whole. First of all, why did the Egyptian Bondage begin with voluntary assimilation instead of forced enslavement, as has happened so often in our history?

The answer lies in a proper understanding of the significance of the Egyptian Exile. We know from the Holy works of our sages that although the sojourn in Egypt occupies a *specific* place in our history as the first *Galut*, it also retains another identity as the archetypal Exile. Far from merely an individual event in Jewish history, the Exile in Egypt is the *prototype of all future exiles*. The Exodus and every detail associated with it, represents a cosmic phenomenon, overlapping the bounds of time and space, encompassing within itself events both past and future.

We may recall the words from Psalms quoted earlier:

"He has made a memorial for his works; the Lord is gracious and full of compassion." The Exodus stands as a permanent reminder of both great events past and what is in store for the future. Micah the prophet is even more explicit in stating the character of the Egyptian Exile and Exodus as the *leitmotif* of all future exiles and redemptions:

"As in the days of your coming forth out of the land of Egypt will I show unto him marvelous things." (Micah 7:15). Clearly, the Redemption of the Messiah is inexorably bound with the Exodus from Egypt.

THE PAST AND THE FUTURE

Although in general, the future Redemption will mirror that of the Exodus from Egypt, in one crucial way they shall differ:

" 'And you shall eat it in haste' (Exodus 12:11). Abba Hanan stated in the name of Rabbi Elazar: 'this haste refers to the haste of the *Shekhinah*. Although there is no proof that this is the proper interpretation, there is nevertheless an allusion that such is the case; for it stated, 'Hark! my beloved! behold, he comes, leaping upon the mountains, skipping upon the hills' (Song of Songs 2:8). Further, it is stated, 'Behold, he stands behind our wall, He looks in through the windows, He peers through the lattice.' Should you perhaps think that in the future Redemption it will be the same, Scripture tells, 'For you shall not go out in haste, neither shall you go out by flight' (Isaiah 52:12)."

Rabbi Eliyahu Dessler illuminates for us this seemingly obscure Midrash. We know, he explains, that the Jews were on an extremely low spiritual plane when they were delivered from

Egypt. Although they did not have it within their own spiritual power to awaken the Divine grace necessary for Redemption, God "hastened" to redeem them before they actually merited the great miracles involved. In the case of the future Great Redemption, we shall have to actuate and awaken the Divine will ourselves. Why this major difference when almost all other aspects of the two redemptions are the same? Rabbi Dessler explains that once the Torah was given to Israel, it becomes our responsibility to initiate Divine sympathy and action. Whereas we once looked to the Almighty for the "handout" of redemption, after Sinai we are expected to "earn our keep."

We can now understand why the Egyptian Exile had to be entered upon willingly and later developed into involuntary servitude. As the archetypal *Galut,* the Exile in Egypt had to encompass within itself every conceivable manner of slavery: physical, cultural and moral and had to include both voluntary enculturation and forced physical labor. The Torah makes clear as early as the Covenant of the Pieces with Abraham that the Egyptians are to be blamed for only one aspect of the Jewish slavery:

"Know of a surety that your seed shall be a stranger in a land that is not theirs *and shall serve them;* and they shall afflict them four hundred years; and also that nation *whom they shall serve* will I judge" (Genesis 15:13).

THE PURPOSE OF THE PLAGUES

The Torah stresses that the punishment is to come upon the Egyptians because they caused the Jews to *serve,* not because they *afflicted* them. Reason would seem to dictate the exact opposite. In blaming the Egyptians for only the enslavement of the Jews, but not for their affliction, the Torah is telling us an important fact about the entire story of the Exile and Exodus in Egypt. The affliction of the Jews, as well as all the miracles done in their behalf, were brought by God to awaken the Jews to the peril of their moral condition and to illuminate the decadence and ugliness of the Egyptian way of life. Thus, each of the plagues accentuated a different aspect of Egyptian depravity. Far from simply being punishments for their misdeeds, the Ten Plagues focused a radiant bright light on the tenebrous Egyptian culture from which the Jews had to be liberated. The redemption from bondage to false ideas and immoral practices is not as easily accomplished as the breaking of physical chains and shackles. This liberation is the truly great story of the Exodus, which is sadly so often overlooked.

The Exodus, because of its archetypal nature, maintains an intricate relationship with the entire history of man. As the Jewish people rose through the forty-nine levels of uncleanliness and slowly gained the forty-nine degrees of understanding, they atoned not only for their own sins, but also for the specific and characteristic wrongdoings of each generation previous to their own.

This relationship with the history of mankind cause the Ten Plagues to take on yet another identity. They stand as an eloquent and detailed compendium of moral teachings, serving as a guide to the past follies of mankind and how to avoid the same pitfalls in the future. Therefore, before we look at the plagues themselves, we must review the history of mankind up to the Age of the Exodus and discover the essence of each generation's shortcomings. Once we understand what went wrong in each of the generations of antiquity, we will be able to understand the corresponding lesson within the Ten Plagues.

Chapter II

THE PATTERN OF HISTORY

ADAM AND EVE

It is absolutely impossible for us to visualize what Adam and Eve were like before their sin. Their physical beauty and stature, their extraordinary wisdom, their literally God-like images are beyond our wildest imagination. Even after the sin, Adam and Eve were great and pious individuals, to whom their progeny continued to come for many centuries to obtain sage advice and counsel. Yet, although we can never totally comprehend their actions and motives, since the Torah relates to us the story of their Fall, it is incumbent upon us to study the passages and commentaries that deal with that cataclysmic event. A primary rule concerning the Torah is that it was given to men, not to angels, and that all within its pages directly concerns and affects the entire human race. It is in this spirit that we must examine the circumstances of that first great transgression.

The eating of the forbidden fruit involved two distinct delictions against the Divine will:

1). disobedience of the Divine edict not to eat of the fruit of the tree of knowledge. This act was the prototype of all future desires.

2). When the Lord asked Adam, "where are you," and he answered, "I was afraid because I was naked," there was a slight implication that the Divine plan was not perfect. In asserting the recognition of nakedness, Adam seemed to imply that God had not completely provided for His work, and that man must improve upon creation. Although Adam, even in his sin, was an extremely righteous man, nevertheless there was a nuance in his words of nonacceptance of the Divine plan. Thus, the Talmud (Sanhedrin 38b) uses extremely strong words to criticize Adam for his transgression.

As an initial example of how the generation of the Exodus was taught by God to atone for the sins of previous generations, we are now in a position to examine the relationship of the manna from heaven to Adam's sin.

THE JEWS TRUST IN GOD

We learned from the Zohar that God literally had to pluck the Jews out of Egypt before they sank to the fatal fiftieth level of uncleanliness. One by-product of this haste was that the Jewish people had not made any adequate preparations for the long journey ahead. The Torah bears witness that "they were thrust out of Egypt and could not tarry, neither had they prepared for themselves any nourishment" (Exodus 12:39). Upon this verse, Rashi comments:

"This passage comes to praise Israel, for they did not say, 'how can we go out into the desert without sustenance,' but they believed and went. Concerning this great display of faith, the prophet declares, 'I remember for you the affection of your youth, the love of your espousals; how you went after Me in the wilderness in a land that was not sown' (Jeremiah 2:2). The reward for that complete trust in the Almighty is mentioned in the very next verse: 'Israel is the Lord's hallowed portion, His first fruits of the increase; All that devour him shall be held guilty, evil shall come upon them, says the Lord.' "

Since the Jews, pitiful as was their moral condition, showed such an unflinching trust in the Lord, God granted them the greatest manifestation of Divine Providence in the history of mankind, the manna.

THE MANNA

The manna fell each day, providing sustenance for that day only (Exodus 16:19). If anyone left some over for the next day, it would become wormy and inedible. This process, repeated daily for almost forty years, "turned the minds and hearts of the Jewish nation toward heaven for sustenance" (Yumah 76a). Every Friday, the manna fell in double portion to teach the lesson of the Sabbath. So it was that in the unique desert university which the Jews attended for forty years, the manna was a constant lesson in trust and belief in God. Children grew up literally looking to heaven for nourishment and the cumulative effect developed a nation that trusted totally in God.

The lesson of manna also corresponded to the underlying causes of Adam and Eve's sin:

1). DESIRE—The manna, by providing each partaker with whatever taste he craved, showed that man's desire can be contained within a shapeless wafer, if he only has the moral strength to comply with the will of God.

2). THE PERFECTION OF THE DIVINE PLAN— The manna provided daily proof that God, in His Providence, provides for all the needs of mankind. By totally believing in Him, as the Jews had done when they left Egypt without preparing for the rigors of the desert, the Jewish nation merited the Personal care that the Lord provided for them during their forty years in the desert.

When Moses, before his death, reviewed for the Jews their glorious history, he capsulized for them the lesson of the manna:

"And He afflicted you, and suffered you to hunger, and fed you with manna, which you knew not, neither did your fathers know; that He might make you know that man does not live by bread only, but by every thing that proceeds out of the mouth of the Lord does man live" (Deuteronomy 8:3).

Moses stressed to the generation that had lived in the desert that the manna was a *once in history* phenomenon, provided to teach a specific eternal lesson. This lesson, of reliance upon the Lord for one's physical needs, atoned for Adam's sin of tampering with the Creator's plan and provided the nation of Israel with the primary trait they would need for the centuries ahead: absolute trust in the Lord.

Once the manna had taught the Jewish people how to atone for Adam's sin, it began to elevate them even higher in righteousness. The manna provided the first and only time in man's history when he received a daily moral judgment and retribution for his actions. The *tzadikim,* the very righteous, awoke to find the manna on their doorsteps; the average person had to go out and gather his food. The wicked found that a long and hard search was necessary for their daily bread. If someone truly repented for his sins and became a *baal teshuvah,* the change was immediately recognized by the manna. Thus, after their desires had been regulated and they had learned trust in the Lord, the Jewish people went on to improve and elevate themselves daily, purifying and cleansing the souls that had been sullied in Egypt, until they would be ready to enter the Holy Land.

Since the manna is so intimately involved in the atonement for Adam's sin, which is mankind's concern until the coming of the Messiah, later generations, too, have not forgotten the manna. Moses instructed Aaron to save a bit of the manna, always to be kept near the Holy Ark as a remembrance (Exodus 16:33) and Jeremiah later had occasion to use it to reinstill the people's trust in God (Jeremiah 2:31). We, too, recite daily the Biblical portion concerning the manna to remind ourselves of that great lesson our ancestor's received in *emunah* and *bitahon*, the proper trust in the Lord.

THE SABBATH

The time of the greatest manifestation of the heavenly quality of the manna, the Sabbath, when none fell because a double portion had been prepared on Friday, is celebrated by placing a double Halah on the Sabbath table. Even the covering of the Halah represents the layer of dew which protected the manna after it fell. Thus, the Sabbath, which provides a taste of the World to Come, the world of Adam before the Sin, is replete with reminders of the manna from heaven. As we rest from the burdens of the mundane world, as we return in time to Pre-Fall Innocence, our Sabbath Halah becomes manna and our cloth covering, a layer of heavenly dew. Just as in the Torah, Adam's sin is not mentioned until after the chapter on the Sabbath, although the sin itself occurred on Friday, we, too, *do not recognize* the sinful state on the Holy Sabbath. Having learned from the manna how to atone for Adam's sin, we rejoice *in innocence* and exaltation in the gift of the Sabbath.

CAIN

Keeping in mind the fundamental thought that it is impossible for us to have a clear concept of the true nature of Biblical personalities, we now proceed to examine the age of Cain. Even after he murdered his brother, Cain continued Adam's sin of seeking to improve upon the Creator's plan. After he was sentenced to be a wanderer (Genesis 4:12), Cain built a city "with the same name as his son Enoch" (Ibid. 4:17). Commentators on the Torah note that the word *boneh*—build, not *banah*—built, is used in describing Cain's architectural work. The *Clei Yakar* explains the use of the present tense in the following way: "People like Cain, who are obsessed with the ownership of things in this world, are never satisfied with what they already possess. Their lives are ruled by their addiction to the accumulation of material objects and they can experience no rest, for they must constantly run to and fro in a never-ending vicious cycle."

Cain's activities extended and deepened Adam's sin of attempting to "add" to and improve upon the already perfect creation, and thus drove mankind further toward destruction. Another indication of Cain's weltanschauung may be seen in the verse concerning the naming of his city. The Torah does not say that he named his city *for* his son Enoch, but *as* his son Enoch, showing that to Cain, the brick, mortar and wood of a city were equally as valuable as the life of his son; thus he gave them both the same name. With Cain, begins mankind's sad and tragic search for happiness *outside* the will and plan of the Creator. As we shall see, succeeding generations until the Flood merely refined and expanded this calamitous quest.

LAMECH

Cain's passion had been for the acquisition of material things. Lamech descended even further into corporeal lust. The Torah relates that "Lamech took unto him two wives; the name of one was Adah, and the name of the other Zillah" (Genesis 4:19). Rashi comments in the name of the Midrash:

"Such was the way of the generation of the Flood. They would take two women, one for childbearing and one for physical pleasure. They would sterilize the one whose purpose was pleasure and feed, clothe and pamper her accordingly. The other one performed her function and was abandoned and rejected as if she were in truth a widow."

This development represented a great perversion of the Creator's design for the world. When woman was created, the Lord had said, "It is not good that man should be alone; I will make him a help-meet for him" (Genesis 2:18). As shown very often in the Talmud, God's plan was for man to be able to share his joys and sorrows, pleasures and tribulations, successes and failures with his partner in life. Lamech took this Divine plan and rent it asunder. He made of Woman an object, rather than the person in her own right God meant her to be.

With this iniquitous innovation, began the era of sexual depravity which led to the ultimate condition when "all flesh had corrupted their way upon the earth" (Genesis 6:12), a phenomenon that could only end in the destruction of the greater part of the world. It was almost inevitable that such an arrangement as Lamech's should give birth to a son like Jabal.

JABAL

All that the Torah tells us about Jabal is that he was "the father of such as dwell in tents and have cattle" (Genesis 4:20). Rashi elaborates in the name of the Midrash: "He built houses for the sake of idol-worship." Another Midrash gives even more details: "With Jabal began the evil inclination to worship idols. At first, this diathesis was insignificant, for Jabal did not dare to worship his idol openly. He hid the idol beneath his tent and would pray to it in secret." Thus mankind took the fateful first step toward totally forgetting the true Creator, although with Jabal the inclination remained in private. His brother Jubal changed even that.

JUBAL

The Torah tells us that Jubal was "the father of all such as handle the harp and pipe" (Genesis 4:21). Jubal's proclivities brought mankind into an entirely new sphere. No longer did Lamech's lust for physical pleasure and carnality dominate man's strivings. Jubal's soul, the soul of the poet and artist, yearned for abstract beauty, for the indefinable sublime. His sensitive ear heard the strains of the universal music, but then he made the fatal error of his ancestors. Instead of *adding* his voice to the universal harmony King David described when he sang, "Let every thing that has breath praise the Lord" (Psalms 150:6), Jabal sought to *replace* this beautiful symphony. Like his predecessors, Jabal did not realize that the Creator had fashioned a perfect universe in which man occupied a special and honored place. Man's task is to reflect the "image of God" in which he was fashioned, not to attempt to supplant the Creator's plan. Ironically, it was the brother with the greater sensitivity, the deeper perceptions and feelings that spread the arrant doctrine of idolatry. While Jabal had worshipped his idol privately, Jubal did so publicly. Once the way was paved for open disobedience to the Lord, it was only one short step to the murderous age of Tubal-cain.

TUBAL-CAIN

The Torah tells us that Tubal-cain was "the forger of every cutting instrument of brass and iron" (Genesis 4:22). Our holy commentators point out that Tubal-cain is not called the *father* of his craft as were his predecessors, for in truth he simply improved and extended the craft of his grandfather, Cain, the profession of murder. The Midrash relates to us the tragic end of these violent generations:

"Lamech was blind in both eyes, but would nevertheless go out to hunt every day, led by his young son Tubal-cain. Lamech would hold the bow in his hand and when Tubal-cain spied an animal, he would alert his father, who would shoot in the proper direction. One day, when they were in the forest, Cain passed through, and since he was quite far, Tubal-cain mistook him for an animal and gave his father the signal to shoot. When Tubal-cain went to retrieve the carcass, he found that they had erred and informed his father, 'sir, instead of an animal, we have killed a man, and he has a horn in his forehead. Lamech realized that he had killed his grandfather and in his anguish, he clapped his hands together and murdered his son."

So ended the lives of the first two human beings whose very existence was defined in terms of the blood they shed. Now that we have studied the history leading up to their demise, we can see that their tragic end was inevitable. Each generation since Adam had erected a new barrier between the Creator and His beings and things had to soon come to a major crisis or solution.

THE GENERATION OF THE FLOOD

After the catastrophic innovations of the family of Cain, the Torah begins to discuss events in terms of entire generations instead of isolated individuals. This change in emphasis is summed up by the verse, "and it came to pass when men began to multiply on the face of the earth" (Genesis 6:1). This passage may be understood on two levels in portending the coming of the Flood. First of all, the Torah alerts us to the fact that whereas in previous generations, certain *individuals* deviated from the path of the Almighty, with this new generation, iniquity was generally widespread and was not limited to any one person or group.

On a deeper level, the Torah's stress on the population increase refers to each individual *within himself*. We know that Noah, like his predecessors, was an inventor. Noah invented the plow and devised other implements to till the field. Since God had cursed the earth, "thorns also and thistles shall it bring forth to you" (Genesis 3:18), mankind had labored manually and with great hardship, producing only dismal results. Thorns, weeds and thistles were rampant and even a small harvest required hundreds of hours of backbreaking toil.

Noah had hoped to alleviate man's desperate situation with the invention of the plow. Recognizing with alarm the depths to which mankind was sinking, Noah surmised that the problem lay in the total lack of leisure. Noah realized that because of the great difficulty of obtaining food from the earth, there was no time for introspection, no moments of tranquility when man could meditate upon his purpose in the world, not a minute he could call his own to reflect on the grandeur from which he had fallen and how to recover it. Noah hoped to solve all this for mankind by providing the precious gift of time.

But Noah miscalculated. Mankind turned around his magnificent gift by misusing the precious moments it had been granted. Instead of seeking the serenity necessary for Divine guidance, the multitudes utilized their newfound leisure to seek ever more corrupt pleasures and novel entertainments. So great became their iniquity that the Mishnah (Sanhedrin 107b) decrees that the Generation of the Flood has no share in the world to come. Robbery and licentiousness literally became a way of life and even the animal world was corrupted by man's perversion.

We can now understand the second sense of the verse, "when men began to multiply on the face of the earth." Instead of using its free time for moral introspection and *decreasing* its frantic activity, mankind began to *increase*, to "multiply" its useless actions and foolish aspirations. Mankind sought new ways to *spend* the precious time it had been given, like one who has been given a tremendous sum of money and only a limited amount of time in which to enjoy it. Sadly, man did not realize that *time was the gift itself*, not to be wantonly spent, but to be cherished and savored.

Only one man, Noah himself, spent his time properly and profitably, increasing his own inner self. The Torah testifies that "these are the generations of Noah. Noah was in his generations a man righteous and whole-hearted; Noah walked with God" (Genesis 6:9). As Rashi comments, "the main progeny of the righteous are their good deeds." Noah gave birth each day to a new man—a *tzadik*—who walked closer and more steadfastly with the Lord than the day before.

But Noah succeeded only in bettering himself. The rest of mankind deteriorated at an ever-increasing rate, sinking deeper and deeper into moral self-destruction. Their sins may be separated into three distinct categories:

1). SEXUAL DEPRAVITY—The Generation of the Flood saw a total breakdown in the accepted fiber of the Family. The Torah testifies that "they took many wives, whomsoever they chose" (Genesis 6:2), which the Midrash understands to include married women and even animals. We can trace the development of the new and strange desires which mankind developed after Adam's sin, in the following way: desire for physical pleasure (the fruit), desire for momentary pleasure (sexual lust) and then the leap into desires which do not even lead to any physical pleasure:

2). POWER—The Torah speaks of the "sons of God" (Genesis 6:2), which Rashi explains to mean "the powerful and the judges." The Generation of the Flood was the first in which men wished to rule over their fellow human beings. The desire for power leads to no recognizable physical pleasure, nor even to the ephemeral satisfaction of carnal sin. The symbolic meaning of the union of the "powerful" with the "daughters of men" is that the leaders of the Generation of the Flood began to combine their twin desires for power and the gratification of lust into fatal coalescence. The result of this combination is called by the Torah *nefilim.* Rashi (Genesis 6:4) explains the root of the word as coming from *nafal*—to fall—and asserts that they "toppled the world." Truly, the tendency to reduce one's personality to a series of drives—for sexual gratification or for raw power—reduced the antediluvian age to an irrational, morally impotent generation, that had nowhere to go but destruction.

3). ROBBERY—Yet, despite the magnitude of all these sins, the Talmud (Sanhedrin 108a) teaches that "their destruction was not finally decreed except for the great sin of their predacity." It was the total disregard for their neighbor's property, for the basic laws of ownership, that caused their ultimate destruction. These three factors which brought about the annihilation of the Generation of the Flood bring to mind the statement of the Mishnah (Avot 4:28), "Envy (they stole, for they wished what belonged to their neighbor), lust and vainglory (the search for power) remove a person from this world." These three arch-sins did indeed remove an entire generation from the face of the earth and later became primary focal points for the great lessons the Jews were to learn during the Ten Plagues.

DOR HAHAFLAGA—THE GENERATION OF THE DIVISION

One of the most extraordinary events in the history of mankind is the building of the Tower of Babel and its cataclysmic aftermath. Almost no aspect of our lives today, from the simplest daily communication, to the most esoteric linguistic analysis, is untouched by the unique circumstances surrounding the *Dor Hahaflaga.*

Before we examine in detail the specific transgressions of the *Dor Hahaflaga,* let us remind ourselves of the proper attitude toward our ancestors in antiquity. Let us remember that we are not dealing with Neanderthal man, nor with any mythic caveman whose only feelings are barbaric and primitive in nature. The Torah rejects such an outlook and recognizes in early man a potentially noble and incredibly intelligent progenitor.

Keeping in mind this all-important proviso about early man we may proceed to examine the strange story of the Tower of Babel. Careful attention to the words the Torah uses in describing their history, reveals to us the inner plans of the leaders of that fateful generation. "They found a *bikah* (plain) . . . and they dwelt there" (Genesis 11:2). The word *bikah* is associated with the word *b'kiah,* which means fission or total disintegration. The Zohar notes that the word "found" is a strange one in relation to the discovery of a plain. The more appropriate word would have been "they spied." Therefore, the Zohar asserts, this refers to their rediscovery of "*an ancient and lost knowledge.*" What was this extraordinary wisdom that they discovered which gave them visions of grandeur, of conquering the heavens and absolute invincibility?

Does it not seem possible that the *Dor Hahaflaga* stumbled upon the long-hidden secret of the atom? What, then, is the science of *b'kiah*—fission—if not the power to split the atom? Let us read a bit further in their plan for an assault on the heavens: "Come, let us make brick, and burn them thoroughly. And they had brick for stone, and slime had they for mortar." Their machinations came in two stages:

1). let us make bricks—in other words, let us utilize our new-found power to invent and fashion new materials that will be able to withstand the unheard-of rigors to which we will put them.

2). let us develop the power of fire—We already know that fire's natural tendency is to rise, and we shall need only to channel it to our desired use. Rabbi Jonathan Eyebeshutz, the early 18th century sage, describes the Tower of Babel as a giant spaceship, with a tremendous fire pushing down so that the Tower would rise to the heavens. Surely, the *Dor Hahaflaga* had achieved great advances in science and technology, perhaps even stumbling upon the secret of nuclear energy.

What did they wish to accomplish with all their efforts? They said to each other, "Let us make a name" (Genesis 11:4). The Talmud (Sanhedrin 109a) explains this to mean that they wished to avoid the necessity of Godliness by piercing the secret of the universe. They wished to conquer the heavens by plucking out the mystery of the cosmos.

Thus, the sin of the *Dor Hahaflaga* was not related to specific desires or temptations as were those of the Generation of the Flood, but went to the very heart of man's existence and purpose on earth. "What is man's relationship with his Creator and with the Knowledge He allows His creatures to gain?" The misdeeds of the *Dor Hahaflaga* are intimately related to many of the problems scientists and philosophers are attempting to solve today.

SCIENCE AND KNOWLEDGE IN THE NUCLEAR AGE

Since Einstein's theories showed their terrifyingly pragmatic applications with the development of the Atomic Bomb, the scientific community has been busily reassessing its moral and social obligations. For the first time, treatises and informal memoirs by scientists have appeared questioning the absolute right of science to blindly pursue all avenues of knowledge, without regard for the consequences to mankind. The Generation of the Division, too, stumbled upon the vast riches of nuclear science. What did they do with their new found knowledge?

The Torah relates to us the aspirations of the *Dor Hahaflaga*:

"Come, let us build us a city, and a tower, with its top in heaven, and let us maks us a name; lest we be scattered abroad upon the face of the whole earth" (Genesis 11:4). The Midrash elaborates upon their inner purposes:

"And it came to pass, as they journeyed from the east (*mikkedem*) (Genesis 11:2). Rabbi Leazer ben Rabbi Simeon interpreted: they betook themselves away from the Ancient (*kadmon*) of the world, saying, 'we refuse to accept either Him or His Divinity.' 'And let us make a name.' The school of Rabbi Ishmael taught, *Shem*—a name—means nothing else but an idol."

THE IDOLATRY OF SCIENCE

What, in fact, was the idol of the *Dor Hahaflaga?* As we have seen in our study, *THEIR IDOL WAS SCIENCE.* In ignoring the "Ancient," they placed all their trust in scientific knowledge. Like Jubal, who failed to realize that Art is a *means* towards serving the Lord, and not an end in itself, the Generation of the Division made Science into the final authority, without regulation or moral focus. Thus, to them, even the Flood, the greatest manifestation of Divine retribution and chastisement in man's history, was explained away as a regularly recurring physical phenomenon: "They said, 'once in one thousand six hundred and fifty-six years the Firmament totters; therefore let us go and make supports for it.' " As Rabbi Dessler so succinctly points out, "since they did not wish to accept the fact of Divine Providence, they *invented* a scientific law to fit their needs."

The parable Rabbi Dessler quotes in the name of the Maggid of Dubnah is especially appropriate here: "A nobleman was traveling in the forest and spied a farmer sitting with bow and arrow; all around him were targets into which he had with perfect aim shot every arrow. The nobleman asked the farmer, 'how is it that you never miss the target?' Answered the shrewd farmer: 'first I shoot the arrow, and wherever it lands, I draw the circles around it.' " The scientists of the *Dor Hahaflaga* did the same. Instead of searching for the truth and accepting whatever it turned out to be, they sought to reinforce their own preconceived notions and ideas.

In the Nuclear Age, mankind cannot afford to allow science to proceed unregulated and without moral guidance. The stakes are too high and the danger is too great. Just as the *Dor Hahaflaga* had to end in a cataclysm because of their misuse of great knowledge, so our age must seek to carefully moniter the activities of scientists to ensure that their knowledge will be used only for mankind. Just as the military needs civilian supervision, so must those who deal purely in science have an objective and morally watchful eye constantly over them.

THE LESSON OF LANGUAGE

How did the Lord resolve the plans of the *Dor Hahaflaga?* "Come, let us go down and there confound their language, that they may not understand one another's speech" (Genesis 11:7). After this had been accomplished, the Torah testifies that "they left off to build the city" (11:3). Since the Torah does not use the words *v'lo yakhlu*—they could not—but *vayahdelu*—they left off—we must conclude that the builders voluntarily ceased their work. It is apparent that the fragmentation of society by languages taught the leaders of that extraordinary generation a profound lesson. What was it that they learned with the introduction of the new languages?

The answer lies in an understanding of the symbolic significance of language. Modern philosophy has rediscovered and returned to the importance of language to mankind's development. Bertrand Russell, George Edward Moore, Ludwig Wittenstein, Rudolf Carnap and others all thought they had found the key to man's verbal communications. Yet, the *Dor Hahaflaga* learned the answer and it was sufficiently powerful for them to voluntarily cease building the tower.

They learned that in order to discover truth, in order to expand their minds and potentials to the absolute limit, they did not have to attack the heavens. Neither was it necessary to experiment with nuclear power to achieve the result of fulfilling their destinies as the chosen beings of God's creation. The separation of man's universal language into a multitude of different tongues proved a powerful lesson in human potential. Even the most seemingly surface of man's accomplishments—the words he uttered, the language he spoke—was infinitely more complicated than he had thought. The *Dor Hahaflaga* was shown Noam Chomsky's "deep structure," but with a profundity he never dreamt possible.

ABRAHAM

Yet, although the *Dor Hahaflaga* had learned a valuable lesson and willingly ceased the building of their tower, mankind nevertheless soon returned to its folly. Despots such as Nimrod flourished and led mankind deep into the abyss of idolatry and the nether-world of witchcraft. Violence was rampant and immorality spread across the face of the earth. It is no wonder that the Talmud (Sanhedrin 97a) refers to the first two thousand years of history as the years of desolation or nothingness. The positive years, the ages of accomplishment, do not begin until Abraham's fifty-second birthday, the pivotal year when the tide turned for mankind and the long trek back to the Lord was begun.

Maimonides describes for us Abraham's early efforts to rescue mankind from its desperate condition and to convince it of the desipience of its ways:

"As soon as Abraham recognized the truth of the One and Only Creator, he began to question the people of Ur Casdim and to dispute with them. He provided for them compelling proof that they had erred in their ways, shattered their idols and brought them to the true path. He taught them that it is proper only to serve the Creator and to render unto Him offerings so that all may learn the truth of His Divinity. When he had convinced them with the strength of his arguments, the king sought to murder him and a miracle was performed for him and he left for Haran.

After this, he began to proclaim powerfully to all the world that there was but one Lord for all the universe, who should be served by all mankind. And so, he would travel from city to city and from nation to nation, gathering the populace . . . and discussing with each individual according to his own level the great truth which would bring them back to the true and proper path. Eventually, tens of thousands joined him, utterly convinced of the truth he had brought to them, and these people became members of his household. Abraham continued in this great work, instilling in the minds of all with whom he met the great truth of monotheism. In addition, he composed books on this subject and instructed Isaac in all these principles."

Abraham represents mankind's first great step away from the deterioration which progressively took place since Adam's monumental sin. In essence, Abraham is a beginning, not just for the Nation of Israel, but for all of mankind. Until Abraham's 52nd year, mankind lived in a state of *tohu*—nothingness—accomplishing nothing for its Creator, existing only by grace of the Almighty and deteriorating rapidly. With the advent of Abraham, mankind was granted a new chance to redeem itself. Now that the truth was once again shining from the face of a human being, any member of the human race could throw his lot in with Abraham and be saved, or continue down the well-populated road to destruction.

ISAAC

Abraham represents for us that most difficult of stages, the beginning. He set the tone, direction and ideal for the new nation. By going out and bringing the truth to all who would listen, he established the universality of that truth. By practicing absolute kindness and charity to *all* who passed his way, he created a foundation of brotherhood and social generosity for the future nation. Yet, there remained a need to build from within, to create an impenetrable fortress against the moral rigors of the outside world. This had to be accomplished through extraordinary inner meditation and personal sacrifice. This second step in the building of the Jewish nation was accomplished by Isaac.

Let us return for a moment to Maimonides' synopsis of Jewish history. After he has described in detail Abraham's prolific lectures and literary activities, Maimonides devotes but one sentence to the work of Isaac: "And Isaac sat teaching and guiding." Isaac, when he was brought upon the altar ready to be sacrificed had taken on all the laws and characteristics of a *korban olah*—a holy sacrifice. Thus, his essence was not in the performance of great historic and dangerous deeds or in public exhortations to his fellow human beings, but in the deepening and perfection of his own personality. Each one of the Patriarchs added a permanent and irrevocable trait to our national character. Isaac provided the inner strength and determination we have needed to survive the centuries of affliction.

JACOB

The Midrash relates to us an interesting ongoing conversation which the Lord had with the angels until the birth of Jacob:

"When Adam was created, the ministering angels asked

of the Lord, 'is this the one with whose name we praise you (when we say *Blessed is the Lord, the God of Israel*)?' The Lord answered, 'no, he is a thief,' as it is written *And he ate from the tree.* Came Noah and they asked once again, 'is this he?' He answered, 'this one is a drunkard,' as it is stated *and he drank of the wine and was drunken.* Came Abraham and they asked, 'is this him?' He answered, 'he is still a stranger, for he has given birth to Ishmael.' Came Isaac and they inquired, 'is this him?' He said to them, 'He loves my enemy,' as it is stated *And Isaac loved Esau.* When Jacob came and they asked once again, 'is this him?' the Lord replied, 'yes, he is the one.' "

We may learn from this important Midrash an essential fact about Jewish history. With the establishment of the house of Jacob, an era ended and another began. Since all of Jacob's children were righteous and great men in their own right, the nation was now ready to enter upon the dangers of a foreign country and culture. Therefore, it was not until Jacob and his entire household were complete that the Lord allowed the Jews to make their hazardous but necessary journey down to Egypt.

ISRAEL

The moral perfection which Jacob reached is symbolized eternally in his agnominal designation as *Israel,* the name by which the nation at large would be called. The story of his renaming is a familiar one:

"And Jacob was left alone; and there wrestled with him a man until the breaking of the day . . . And he said unto him: 'What is your name?' And he said: 'Jacob.' And he said: 'your name shall be called no more Jacob, but Israel; for you have striven with God and men and have prevailed.' "

Later, the Lord Himself confirmed this name change:

"And God said unto him: 'Your name is Jacob: Your

name shall not be called any more Jacob, but Israel shall be your name;' And He called his name Israel."

Rashi teaches us in the name of the Midrash that the "man" who wrestled with Jacob was in fact "the heavenly agent of Esau." In this light, the name *Israel* takes on the added significance of "he who can survive in a world replete with enemies."

Abraham had established a firm foundation for the house of Israel. Isaac had fortified the edifice from within with holiness and purity, the essential building blocks of permanence. Jacob provided the final touch, that of impenetrability. Jacob's designation as Israel represents the readiness of the Jewish nation to meet its enemies.

MOSES AND AARON

The Patriarchs had left the nation of Israel with an assurance of indestructability. Abraham's universal love, Isaac's piety and Jacob's moral strength had combined to forge a nation that *could not* be destroyed. Yet, it remained necessary to mold the moral fiber of that nation into a "kingdom of priests." It was not enough that Israel would continue to exist under all circumstances. That had been assured by the Patriarchs. But survival was not enough. The Nation of Israel, which was to provide the model to the world of the ideal ethical and moral life, had to be taught the proper way to live that life, even in a world of disbelief and evil. This became the work of the Greatest of Priests, Aaron.

We know that Aaron was involved in his holy work of purification even before Moses became the leader of Israel:

"Is there not Aaron your brother the Levite? I know that he can speak well" (Exodus 4:14).

Indeed, we know from the Midrash, that the entire reason for Moses' reluctance at accepting the mission God had thrust upon him was his deep respect for his brother, who had for a long time acted as shepherd to the Jewish people:

"The Rabbis say that one must not think that Moses refused to go, he only wished to pay respect to Aaron; for Moses said: 'Before I arose, my brother Aaron prophesied to them in Egypt for eighty years, as it is written: 'I made myself known unto them in the land of Egypt' (Ezekiel 20:5). Whence do we know that it was Aaron who prophesied? For so it says: 'And there came a man of God unto Eli, and said unto him: Thus says the Lord: Did I reveal myself unto the house of your father, when they were in Egypt in bondage to Pharaoh's house? And did I choose him out of all the tribes of Israel to be My priest?' (I Samuel 2:27-28). Moses thought: If I now trespass upon my brother's domain, he will be vexed. On this account he was reluctant to go."

We come now to a crucial and delicate question. Why indeed was Moses chosen over Aaron to lead the Jews out of Egypt and did Aaron retain any of his responsibilities as former leader of the Jews? We may obtain a clear understanding of the relative roles of Moses and Aaron through a unique interpretation of Rabbi Joseph Rosen, known as the Rogatsover Gaon.

Rabbi Rosen seeks to explain the phenomena of Moses' apparent speech defect and his need for Aaron as an interpreter, when juxtaposed with Moses' great accomplishments in virtually every field of human endeavor. The paradox is blatant and demands explanation. Why did Moses seem to have difficulty only in speaking with Pharaoh? We know that later he constantly taught and instructed the Jewish people and there is no mention of his verbal difficulty! Thereupon Rabbi Rosen reveals to us a fundamental factor in the personality of *Moshe Rabeinu*:

Moses was the living personification of Torah. He had no other existence and could reflect no teaching other than the

Torah. For this reason, because the Torah is not only written, but can be understood fully only in Hebrew, *Moses could speak no other language.* Although Moses could, of course, speak the Egyptian language when growing up in Pharaoh's palace, from the moment he accepted the Torah as the lifeblood of his being, his entire personality was changed. He was no longer Moses, who had lived here or there, or even Moses son of Amram, but Moses the living embodiment of Torah. And since, according to the laws of prophecy, a prophet cannot speak through an interpreter, Aaron had to serve Moses as "a mouth."

Rabbi Rosen goes even further in his exploration of the quintessential differences between Moses and Aaron. How, and why, did these differences occur? Surely both were absolutely righteous men. Both were totally dedicated to serving the Lord. The answer, the Gaon asserts, lies in the mode of acquiring Torah knowledge. To Aaron, the wisdom and understanding of the Torah were given as a gift. Abraham, too, did not have to labor to understand the difficult portions of the Torah. Only Moses had to persevere and toil over the Torah until finally understanding was granted him.

Although it is infinitely more difficult, acquisition by personal labor has definite advantages over a free gift. The earned reward, having been toiled over and sweated for, becomes a part of the person himself. Instead of remaining without, it becomes an integral, inseparable component of the human being. And so it was with Moses our Teacher.

While Moses was perfecting his knowledge and understanding of the Torah, Aaron was perfecting his *human* qualities of kindness, mercy and love of fellow man:

"Every day, Aaron would wander to and fro in the Jewish camp and attempt to make peace between two people who were arguing and between husbands and wives who had quarreled. Often, he would be particularly friendly to a man who was known to be a sinner. The next day, when the man was about to return to his evil ways, he would think to himself, 'how will I be able to look Aaron in the eye tomorrow when he greets me in his usual pleasant way?' "

"When Aaron saw two people arguing with each other, he would go to one of them and say, 'my son, do you know what your friend is doing? He is eating his heart out for anguish and tearing at his clothing for mortification, saying: 'How will I be able to face my friend? I am so ashamed that I have wronged him.' Aaron would remain with him until he had removed the anger from the man's heart; then he would proceed to do the same with the other party in the dispute. When the two met, they were already reconciled and now appreciated each other's friendship and good qualities."

While Moses was the personification of Torah, Aaron was the personification of kindness toward fellow man. And here we come to the crucial transition from Aaron to Moses as leader of Israel. It was Aaron who molded the Jewish people into *one nation*, who loved and were responsible for one another, who gave up their lives for each other and were ready to face a hostile world as one unified entity.

But the most important step was yet to come. The exile in Egypt was specifically designed to test Israel against the incarnation of knowledge in the physical world; this was the nation of Egypt epitomized by its leader, Pharaoh. Egypt was not only the center of learning in the world, but the embodiment of all acquired human knowledge; the only force that could counteract such an influence would be the *intellectual* and spiritual force of the Torah. Only Moses was able to utilize and transform this force into the weapon which would release the Jews from bondage, while gaining the valuable spiritual lessons they would need for the long journey ahead. Therefore it was Moses, and Moses alone, who could lead the Jews through the forty-nine levels of impurity into the holy world of the Torah.

Chapter III

PRELUDE TO THE PLAGUES

PHARAOH AND THE SERPENT

Although the Egyptians themselves did not make a conscious effort to entice the Jews into imitating their immoral life-style, the same cannot be said of Pharaoh. The Prophet Ezekiel is explicit in condemning Pharaoh for his blasphemy as well as for his treatment of the Jews:

"Thus says the Lord God: Behold I am against you, Pharaoh King of Egypt, the great dragon that lies in the midst of his rivers, that has said: My river is my own and I have made it for myself. . . . And the land of Egypt shall be desolate and waste, and they shall know that I am the Lord; because he has said: The river is mine and I have made it" (Chapter 29).

Pharaoh's characterization as a dragon or serpent is crucial to an understanding of the events immediately preceding the Ten Plagues. Although the Lord instructed Moses to prepare three signs to present to Pharaoh, in fact he showed only one, the transformation of the rod into a serpent. Since, as we know, Pharaoh's sorcerers soon duplicated this feat, it is difficult to understand why God chose this particular sign.

Fortunately, the Zohar asks this very question and provides a profound answer:

"Rabbi Hiya asked Rabbi Yossi: 'Surely the Lord knew in advance that the *hartumim* (Pharaoh's sorcerers) could also turn their rods into serpents. Why, then, did He instruct Moses to present this sign to Pharaoh?' He answered, 'because all of mankind's troubles began with the Serpent and the source of Pharaoh's power was with the Serpent.' "

The Zohar teaches us that Ezekiel's characterization of Pharaoh as a serpent is no mere metaphor. Pharaoh's power, sovereignty and puissance are ultimately related to that of the Primeval Serpent. As we shall later see in our examination of the Plague of Darkness, the Serpent was originally created to be the King of the Beasts. His wisdom was only one step removed from that of man:

"And the Serpent was subtle" (Genesis 3:1). After his sin, he was demoted to groveling on the ground as the lowest and most despised of beasts, but his animalistic power, the churning lusts within his being, the enormous potential for evil creativity, remained in the world, ready to be tapped by the diabolical powers of evil.

Throughout history, evil geniuses have made sinister use of this vast dangerous fund of knowledge. Perhaps that art is known today by other names, but those who penetrated the secret of how to control man's nature have always been able to decide the destiny of individuals, nations and empires.

Pharaoh is the prime example of this type of personality. Having achieved a pinnacle of knowledge in the occult arts, well versed in the science of human temptation and submission, Pharaoh molded his people into the image *he* sought to create. Egypt under his tutelage soon earned a well-deserved reputation for decadence and evil unparalleled in the rest of the world.

In presenting to Pharaoh the sign of the serpent, Moses was giving him notice that the Lord knew the source of his evil power. In having the rod of Aaron swallow up the rods of Pharaoh's *hartumim*, God was giving Pharaoh and the Egyptian people notice that the power of the Serpent had been conquered, no matter what form it took, and the time had come to repent and discard its influence.

On another level, the entire Exodus from Egypt represents mankind's emancipation from the jaws of the Primordial Serpent. The Torah specifically utilizes the imagery of rescue from the depths, in describing the Exodus:

"Or has God assayed to go and take Him a nation *from the midst of another nation*" (Deuteronomy 4:34). Israel had been swallowed up by the Serpent of Pharaoh, the entire Egyptian nation having become a serpentine vehicle for Pharaoh's perversions, and they had to be liberated both physically and spiritually from its grasp.

THE NILE

The Nile symbolized for the Egyptians all their achievements in the intellectual world. It is no wonder that Pharaoh claimed for himself the creation of their sacred river:

"My river is mine own, and I have made it for myself" (Ezekiel 29:3). The Egyptians looked to the Nile, not only as the source of all irrigation, but as the embodiment of their entire culture. When Pharaoh claimed the Nile for himself, he was announcing to the world that he alone had structured and designed the Egyptian civilization. The Zohar tells us quite explicitly that while the Nile itself symbolizes the general weltanschauung of Egypt under Pharaoh, the fish, frogs etc. which inhabit the water correspond to the customs, thoughts, demeanor and general tone of the daily life of Egypt's populace.

Just as Moses was the direct antitheses of Pharaoh, so is the Torah the diametrical opposite of Egyptian culture as symbolized by the Nile. Our sages in the Talmud (Taanit 7a) tell us:

"Why are the words of the Torah likened unto water—as it is written, 'Ho, everyone that thirsts come for water?' This is to teach you that just as water flows from a higher level to a lower, so too the words of the Torah endure only with him who is meekminded."

The medium is the same; both the Nile and the Torah are associated with water. But the similarity ends there. Pharaoh uses the Nile to further his boastful personal goals. The Torah will flourish only within one who is modest. Pharaoh, having attained a certain stature and wisdom, exults and flaunts his so-called creativity:

"My river is my own; I have made it for myself." Moses, at the peak of his career, is considered the most modest man on earth:

"Now the man Moses was very meek, above all the men that were upon the face of the earth" (Numbers 12:3).

Thus, many of the Plagues center about the Nile. Either the Nile itself is smitten, its creatures attack Egypt, or the warning of the next Plague is made at her shores. The Nile, symbolizing both the tremendous potential and the terrible actuality of Egyptian civilization, is the ideal target for the didacticism of the Plagues. If the Nile had remained untouched and unaffected by the Plagues, the average Egyptian could not have learned the lessons in store for him. Once the Nile was shown to be vulnerable and indeed, at one point, odious, the Egyptians could be weaned from their disasterous life-styles and returned to normalcy. As the Children of Israel watched the Nile—in other words—the culture of Egypt—destroyed from within, they were permanently immunized against the rigors of the centuries of Exile ahead. Undoubtedly there would be dangers; inevitably there would be temptations; unfortunately, there would be capitulations. But never after the lessons of Egypt would the entire Jewish nation come to the brink of total assimilation. The Nile had been conquered by its antithesis, the power of Torah, and Pharaoh had been overcome by his counterpart in the world of holiness, *Moshe Rabeinu.* The battle was not a momentary one. Just as *Galut Mitzrayim* was the prototype of all exiles, so were the lessons learned from the Exodus eternal ones, which have served the Jews well throughout the centuries. In the ensuing pages all we can do is to verbalize some of the truths that have been known and *utilized* by the nation of Israel throughout the ages.

Chapter IV

BLOOD: THE LESSON OF KNOWLEDGE

As the first of the Plagues, *Blood* occupies a dual place in the Exodus. It served as an ominous omen to the Egyptians of the extraordinary events to come and also corresponded to the first sin of the *Dor Hahaflaga*, the misuse of knowledge. Even upon a superficial reading of the Passover Hagadah, we are struck by the singularity of the Plague of Blood:

THE "WONDERFUL" PLAGUE

" 'And the Lord brought us forth out of Egypt with a mighty hand, and with an outstretched arm, and with great awe, and with signs and with wonders' (Deuteronomy 26:8). *With a mighty hand* refers to Pestilence. . . . *With an outstretched arm* refers to the sword. . . . *With great awe* refers to Divine Revelation. . . . *With signs* refers to the staff. . . . *With wonders* refers to the Plague of Blood."

The only two plagues which are mentioned by name are Blood and Pestilence. Some commentators see in "the sword" a reference to the Plague of the Firstborn, but even this is equivocal. We shall examine the reason for the special mention of Pestilence in a later chapter. Let us now determine the special significance of the Plague of Blood.

Why should only the Plague of Blood be called a wonder? Surely each of the Plagues was wonderful in its own right and astounded the Egyptian people. The Maharal of Prague provides us with the answer. If we analyze, he says, all the Plagues, we find that they are all within the realm of nature. In each of the plagues, a natural phenomenon is extended, enlarged or modified to castigate the Egyptians. On a smaller scale, frogs, vermin or locusts *do* sometimes intrude upon the world of man. Wild animals occasionally do attack human beings, pestilence is a fairly common menace, hailstorms destroy crops and boils plague the unfortunate victims of skin disease. The darkness brought upon Egypt was a miraculously intense one, but the phenomenon of darkness itself is natural. Even the dying of the firstborn could—to the very stubborn—be construed as resulting from natural causes.

Only the Plague of Blood was a *wonder* in the sense that it clearly transcended the bounds of nature. There is no natural explanation for water turning to blood and the Egyptians had none. While most of the Plagues were kept within the bounds of nature to show that the rules of the universe can and should be used for the Divine will, the first of the Plagues had to establish God's transcendence over nature and His ability to structure a new world with new laws at His will.

THE NATURAL AND THE SUPERNATURAL

The Maharal points out another important aspect of the Plague of Blood, which also would classify it as a *wonder*. The Midrash relates the various machinations of the Egyptians to obtain water despite the Plague:

"Rabbi Abin the Levite, the son of Rabbi said: The Israelites became wealthy from the Plague of Blood. How was that? If an Egyptian and an Israelite were in one house where there was a barrel full of water, and the Egyptians went to fill a pitcher therefrom, he would discover that it contained blood; however, the Israelite would drink water from the same barrel. When the Egyptian said to him, 'Give me some water with your own hand,' and he gave it to him, it still became blood. Even if he said to him, 'Let us both drink from one vessel,' the Israelite would drink water and the Egyptian blood. It was only when the Egyptian had bought it from the Israelite for money that he was able to drink water, and this is how the Israelites became rich."

As the Maharal notes, this phenomenon represents that most extraordinary of manifestations: the intermingling of the natural with the supernatural, the usual with the miraculous. As we know, the Egyptians were no strangers to the supernatural. In a society where the basis of education was the occult, the "miraculous" became almost commonplace. The Plague of Blood alerted the Egyptians to the fact that this was no ordinary conjuration, as they were perhaps accustomed to. They realized immediately that there were Forces at work of which they had no knowledge and they began to heed, even if Pharaoh did not.

Essentially, we find, that the Plague of Blood was Divinely designed to accomplish three monumental tasks:

1). As the first of the Plagues, it had to establish the Divine character of the retribution upon Egypt. It had to be abundantly clear that failure to respond affirmatively to the demands of the Lord would bring catastrophe upon Pharaoh and his people. This, we have seen, was accomplished through the *wonderful* mixture of the miraculous and natural within the Plague itself.

2). The Plague of Blood had to demonstrate the didactic nature of the coming punishments upon Egypt. It was essential that Pharaoh perceive that not only were he and his people being punished for past misdeeds, but that mankind in general was being taught an eternal lesson.

3). The Plague of Blood was brought specifically to atone for the *Dor Hahaflaga's* sin of misusing the great knowledge they had gained.

We have already seen how the first of these tasks was accomplished. How did the Plague of Blood prove irrefutably the moralistic nature of the Plagues and what did it have to do with Egyptian misuse of their accumulated knowledge?

We will recall from Chapter III that the Nile represented the sum total of all Egypt's cultural, scientific and technological achievements. Those who marvel today at the ingenious construction of the pyramids and the mysteries of the Sphinx can well imagine the awe in which their conceivers and architects were held in ancient Egypt. The Nile, seemingly eternal, mystical, powerful, the key to prosperity and source of all irrigation, was considered to be the fount of all life and wisdom, the one immutable in a world of change.

It was thus most fitting that the first plague should be upon the Nile itself. In a deeper vein, any change in the revered Nile was a reflection upon Pharaoh and Egyptian civilization. As we read earlier, Pharaoh himself equated his own divinity with that of the Nile:

"My river is my own, and I have made it for myself" (Ezekiel 29:3). An attack upon the sanctity of the Nile was a direct impugnation of all the Egyptians held sacred. The

Plague of Blood began the irrevocable process of eroding the Egyptian sense of complacency, arousing them from their decadence and providing the nation of Israel with an unforgettable lesson for the difficult millenia ahead.

The warning to Pharaoh concerning the Plague of Blood reveals four stages in the manifestation of the Plague: the water shall turn to blood; the fish in the river shall die; the river shall become foul; the Egyptians shall loathe to drink water from the river. Once we realize that the waters of the Nile, in addition to their literal meaning, are symbolic of Egyptian civilization in general, the four stages of the Plague take on profound additional significance:

1. THE WATERS TURNED TO BLOOD

As the waters of the Nile turned red, the Egyptians were forced to realize that nothing material is permanent, even their sacrosanct river-god, the Nile. As the duration of the Plague was prolonged and the one topic of conversation in every Egyptian home was the sudden transformation of the Nile, the Egyptian people began to question all of their heretofore cherished beliefs, along with their deteriorating belief in the Nile. The first effect of the Plague of Blood was to send doubt into the minds of the Egyptians and to alert the Jews to the coming lessons in becoming the Holy Nation for which they were destined.

Since the purpose of the plague was to demoralize the Egyptians and to begin to raise the Jews out of slavery, the plague had a residual benefit for the Jews as well. We read earlier in the Midrash that it was the Plague of Blood which enriched the Israelites. It is no coincidence that of all the Plagues, it was "Blood" that brought the Jews great wealth. On a primary level, it was the first indication that God was now fulfilling his promise to Abraham of redemption and the first step was the acquisition of great wealth: "And afterwards shall they come out with great substance" (Genesis 25:14). Although this aspect of the redemption could have been left till the end, the Lord wished to raise the Children of Israel from their misery with the very first plague.

In a more subtle way, the Jews' newly acquired wealth served notice to both Jew and Egyptian of the ephemeral quality of earthly riches. It became clear to all that no amount of technology, advanced culture, scientific achievements etc. can stand in the way of the will of the Lord. The *Dor Hahaflaga*, too, had acquired great wisdom and knowledge; yet because they misused their great knowledge, they were dispersed and destroyed.

2. THE FISH IN THE RIVER SHALL DIE

A popular photograph in books on ecology and conservation pictures a hopelessly polluted river with dead fish floating on the water. It was just such a sight, multiplied ten thousand-fold that greeted the Egyptians during the Plague of Blood. The Egyptians could not help but begin to contemplate the impotence of their revered Nile against the forces of God.

On the symbolic level, the fish of the Nile represent those who are nourished most directly by the culture of Egypt. The "fish" of Egyptian culture might be the magicians or the upper-class court favorites who exist constantly in the aura of Egyptian decadence and immorality. The large-scale death of the fish was for them an ominous indication of their own future, if they continued in their depraved ways.

3. THE RIVER SHALL BECOME FOUL

As we shall later learn in detail, the Plague of Hail atoned for misuse of the sense of hearing and the Plague of Locusts atoned for misuse of the sense of sight. Yet, the sense of smell, most capable of evoking feelings of revulsion and repugnance, was brought into play as early as the Plague of Blood to shock the Egyptians out of their moral stupor.

Both Egyptian and Jew were suddenly enabled to think objectively about the Nile. Could this source of effluvia indeed be a deity? Could anything as fetid as the decaying and noisome Nile indeed be divine? Clearly and manifestly not. As soon as the Nile lost its majesty and power, the way was open for new ideas, for freshness, for truth. Yet, by watching the Egyptians, the Jews were to learn one more valuable lesson in surviving the future dangers of assimilation and enculturation:

4. THE EGYPTIANS SHALL LOATHE TO DRINK WATER FROM THE RIVER

Moses had warned Pharaoh that after the first three stages, the Egyptians would "loathe to drink water from the river." Yet, when the Plague actually struck, the Torah relates that "the Egyptians *could not* drink water from the river." Apparently, despite all the proof that the waters of the Nile were not fit for consumption—that the Egyptian culture, life-style and civilization had been proven untenable—still the Egyptians hovered helplessly about the Nile, not knowing how to restructure their lives. So desperate were they that "all the Egyptians dug round the river for water to drink."

Let us picture the scene, for it is a pivotal one in Jewish history. With the smiting of the Nile, the Jews were beginning to regain their self-respect, and their new financial independence allowed them to start shedding the trappings of Egyptian culture which they had acquired. Before them, the Egyptians were groveling humiliatingly in the dirt near the Nile, trying to find a bit of fresh water. Not willing or unable to release themselves from the spell of Nile-worship, they continued their futile search "round about the river."

As they watched this awesome scene, the Jewish people were reminded of the *Dor Hahaflaga* when false theories were also clung to and excuses and explanations were made for their shortcomings. As we learned earlier, the *Dor Hahaflaga* made up fact to fit theory and not vice versa. So, too, the Egyptians could not pry themselves free of their disproven beliefs and calamitous customs and declare their intellectual freedom.

THE POLLUTED RIVER OF KNOWLEDGE

The Jews learned a great lesson from this re-enactment of the *Dor Hahaflaga's* folly. Men will cling to that with which they are familiar long after its reliability has been disproven. Our own generation provides graphic, if tragic, proof of this trait within human nature. The liberalization of moral standards, the permissiveness in relations between the sexes, the disasterous experimentation with sacred human bodies etc., have all begun on college campuses, often with the sanction of prominent professors and so-called "guidance" specialists.

Yet, despite the obviously catastrophic results of these innovations, society returns again and again to the morally empty universities for the answers to the problems of our youth. The pure clean water of knowledge has turned into the blood of violence, the death from drugs and their after-effects have been terrifyingly high, the river of knowledge has grown foul with the stench of immorality and licentiousness, the common people have recognized the looseness for what it is and rejected the "waters" of liberalized morality. Yet, where do the intellectuals look for their salvation? From where do they expect to draw the purified waters of renewed morality and ethics? They continue to dig "round about the river for water to drink."

Instead of recognizing the total folly of their ways and

searching for traditional morality, rooted in religion and respect, they futilely concoct new theories and conjectures, all hovering about the same dead fulsome base.

The Plague of Blood provided an extraordinary and "wonderful" lesson in the proper appreciation and use of knowledge. It stressed the scientific method of fitting theory to fact, yet it rejected scientific dogma as the last wellspring of the confirmed atheist. Truly, the Plague of Blood provided a firm foundation upon which the rest of the plagues could be presented.

Chapter V

FROGS: THE LESSON OF GOVERNMENT

Perhaps the most complicated of all the Plagues was the Plague of Frogs. While *Blood* set the tone and pattern for the coming events, the Plague of Frogs was the first to demonstrate in intricate detail the folly and insipience of the Egyptian way of life. The Torah meticulously details every aspect of the Plague of Frogs:

". . . I will smite your borders with frogs. And the river shall swarm with frogs, which shall go up and come into your house, and into your bed-chamber, and upon your bed, and into the house of your servants, and upon your people, and into your ovens, and into your kneading-troughs. And the frogs shall come up both upon you, and upon your people and upon all your servants."

As with every narrative in the Torah, particular details are recounted only to further important teachings and precepts, and certainly not to enhance any literary aims. We may therefore carefully note that the frogs were to come upon Egypt in two waves. The first attack was to take place against seven different elements of the society, represented by the "house," "bed-chamber," "bed," etc. The second wave would attack specifically "you, your people and all your servants." Furthermore, the first attack would come when the frogs would "go up and come" into the Egyptian houses etc. The second attack would be more direct when the frogs would simply "come up." Obviously, the Torah, whose origin is Divine, chooses its words very carefully and we may find vast treasures of wisdom hidden in its choice of phrasing and vocabulary.

Our first object must be to understand the significance of the word *tzfardeah*, the Hebrew word for frog. If we divide the word into its two syllables—*tzfar* and *deah*, we arrive at the meaning "to chase away" (*tzfar*) sense or wisdom (*deah*). As we shall see in the case of each of the plagues, the Plague brought by God only accentuated the fact that Egypt was already suffering *morally* from the affliction that God was visiting upon them physically. Thus, the *tzfardeah* indicated in each of the places they suddenly appeared that the *deah*—wisdom in that aspect of life had flown away, like a *tzipor* (bird).

Although virtually every facet of Egyptian life was corrupt and peccant, the Plague of *Tzfardeah* clearly placed the blame at the feet of those responsible for the welfare and guidance of the populace, the government. Each of the targets of the *tzfardeim*—i.e., the creatures who show that reason has flown away—represents a different branch of the *enforcing* bureaus of government.

I. *YOUR HOUSE*—THE HOUSING AUTHORITY

The prime building block of communities remains today, as it was in ancient Egypt, the apartment or private home. Yet, the planning of homes, both structurally and architecturally can have a crucial effect on the quality of life which exists in those homes. Even the gentile prophet Balaam, recognized the extraordinary holiness mirrored in the arrangement of the Jewish dwelling places:

"How goodly are your tents, O Jacob, your dwellings, O Israel!"

The great sage Rashi explains that Balaam was so extraordinarily impressed by the tents of the Jews because "their tents were so placed that the doorways did not face each other, keeping the high standard of modesty set by the laws of the Torah.

It is the function of that branch of government which deals with housing to see that homes are physically safe—fireproof, structurally sound etc.—and *morally* secure. While a secular government may not be able to *legislate* morality, it can and must provide the conditions to make morality favorable. Oftentimes, in ancient Egypt, as in modern society today, The Home was not a structure designed to encourage proper family life. A child who yearns for a moment of privacy must seek for it in the street, where he usually finds something else instead. Sometimes, at the other extreme, The Home, instead of being the living thriving family center it should be, becomes a showcase for the status-conscious adults who inhabit it. Instead of a warm, secure place of refuge, the home often becomes merely a museum to show off the accumulated bagatelles of the parents. It is up to both government and individuals to insure that houses will be homes and that the ideals of family purity can be carried out as the natural pattern of life.

II. *INTO YOUR BED-CHAMBER*— THE MINISTRY OF HEALTH

Since the Torah is never repetitious needlessly, we must ask why it is necessary to mention the bed-chambers, if the Torah is going to tell us that the frogs entered the beds themselves. Surely, it then becomes obvious that the frogs also entered the bedrooms. The answer is that each of the seven "targets" which the Torah mentions represents a totally different segment of society. If we translate the words *hadar mishkavkha* (your bed-chamber) literally, it means "the room where you lie down." We may notice that there is no mention of sleep, which is a natural, healthy phenomenon. Therefore we may conclude that *hadar mishkavkha* refers to the hospital bed or, in general, to the health department of the nation.

If the government of a country is peopled by leaders who truly care about the welfare of their nation, their first thoughts are to establish the best possible health care system for their people. They themselves can always get the best of care; the test of their compassion and sense of brotherhood lies in the anguish they take in the pain of others. In ancient Egypt, as in many corrupt societies, the elite ruling class cared nothing for the well-being of those in the lower social strata and thus allowed an inferior and mercenary medical establishment to rise.

As we shall see later in greater detail in our study of the Plague of Wild Animals, the Egyptian scientific community grew arrogant and dangerously improvident in their experimentation upon both animals and human beings. After decades of *using* fellow humans as controls and objects of experimentation, human life soon lost any viable meaning and large-scale "scientific" murder became prevalent. Thus, the frogs—the symbol of abandoned good senses—attacked the Ministry of

Health, the laboratories etc., with a vengeful force to indicate the sinister wrongs being perpetrated there.

III. *AND UPON YOUR BED*—THE MINISTRY OF CULTURE

The bed has been the symbol of man's sexual impulse since time immemorial. Although this aspect of life should be the most private and indeed sacred, in decadent societies it becomes a matter of public domain and blatant immodesty. As the Plague of Boils indicated in greater detail, Egyptian culture suffered from the same malaise as our modern civilization: a rampant immorality and absolute shamelessness when it came to sexual matters. In this, the catastrophic decisions of two branches of government must be held responsible:

a). THE MARRIAGE AND DIVORCE PROCEDURES

Very often, it becomes obvious that the laws of nature and those of man seem to be at odds. A dramatic instance is the discrepancy between biological maturity and adulthood according to law. In ancient Israel and many other societies which took a pragmatic view of the facts of nature, marriage took place very soon after puberty. In this way, many of the tragedies of our youth today—the dangerous experimentation, the disparity between very real passions and lack of outlet etc.—were avoided. Of course, one can say that times have changed and what once was, is no longer practical. In some cases, that may well be; yet, when no alternative is provided, it is the government that is at fault. It is the responsibility of adults either to grant to biological adults the full rights of chronological adults, or to provide an atmosphere of chastity and morality, wherein temptation will not stare from every street-corner. This brings us to the activities of The Ministry of Culture:

b). THE MINISTRY OF CULTURE

Although there are many factors that combine to create what eventually becomes known as "the national character," the most important of these is what is loosely referred to as "culture." Through the discreet use of government grants, persuasion, public image of leaders etc., a distinct image of a country or people begins to emerge. If the leaders of a nation wish to preserve the vitality and spiritual integrity of their people, they are duty-bound to watch with infinite care the expression of art in their land.

This brings us to the delicate question of censorship. Shall we indeed agree with the pornographers that man is free to peddle whatever he wishes, if a country is to be free! The analogy must be drawn to the man who is arrested or issued a summons for polluting the rivers or streets of his native land. Surely, everyone agrees that freedom in such a case is limited by the constraints of how much one person's freedom infringes upon the freedom of another. Yet, the purveyor of filth does infinitely more harm to the young minds and hearts he sullies than all the land and sea polluters in the world. The ultimate tragedy of many "advanced" civilizations—that of ancient Egypt, and sadly, our own, included—is that the spiritual is not given equal weight with the physical, the moral is cast aside in favor of more attention to the popular mundane, and in the process, it is the people at large who suffer.

The *tzfar-deah* which entered the museums and libraries, the art galleries and theatre houses, indicated to the Egyptians the wisdom they had abandoned in years of building their decadent culture. Like modern newsstands, the Egyptian streets provided available pornography to every seeker and entered the minds of the impressionable young at the most tender of ages. As Egyptian society crumbled before them, the Children of Israel learned an invaluable lesson in the importance of modesty, moral standards and sane and pragmatic marriage laws.

IV. *AND INTO THE HOUSE OF YOUR SERVANTS—*

THE CABINET

The governmental servants who are closest to the king or leader are the members of his official and unofficial cabinet, the advisors and friends upon whom he relies for advice and counsel. These advisors, whose responsibility it had been for decades to assure the defense and security of Egypt, erred in three essential ways:

1). WHO IS THE ENEMY?

For years, the ministers of defense, Pharaoh's closest counsellors had invented a nonexistent enemy: the forces of morality which wished to destroy Egypt. With the advent of the people of Israel, the Jews personified for the Egyptian ministers the fears and qualms they had harbored for so long. Instead of recognizing that the enemy was within, that the forces that could destroy Egypt were embedded in the culture and mores of their demoralized civilization, the Egyptian cabinet projected all of their fears and misgivings upon the people of Israel.

2.) WHAT IS THE NATURE OF MAN?

In assembling their armies, the Egyptian defense ministry gave no thought to the nature of man. They blindly conscripted men and women, morally fit or unfit and thus created their army. Jewish tradition is, of course, extremely strict about who *may* enter the army and who is *allowed* to fight in the wars.

There are strict age limitations and women could not go to war. The first and primary objective was always to preserve the sanctity of the home and the *moral* nature of the armed forces. The Torah, in its wisdom, recognized that one about to risk his life in the most delicate and dangerous of situations must be given the mental freedom from temptations and distractions which could literally kill him. In forming their *immoral* army against spirituality, the Egyptians took no notice of the spiritual nature of man, and sought to eradicate the last vestiges of morality within the human soul.

3). THE ULTIMATE WEAPON

Although in war, weapons play an essential role, yet they are only useful if the soldier has been properly trained in their use. If one who is inexperienced handles a gun, let alone sophisticated and destructive weaponry, the result can be arbitrarily and personally destructive. The Egyptian ministers utilized the ultimate weapon in their wars, the power of human nature. Yet, because they did not bother to try to understand or appreciate it, they perverted whatever they touched and brought destruction upon their land and people.

V. *AND UPON YOUR PEOPLE—* THE MINISTRY OF EDUCATION

The ultimate responsibility for "creation" of a people rests upon the educators. It is the teachers, authors of textbooks, guidance counsellors, administrators etc., who mold the minds of children into the kinds of adults they wish them to be. The educator bears the awesome responsibility for creating or ruining a human being, for developing a healthy righteous individual or a perverted antisocial outcast, who is no good to himself or his fellow human beings. What was the Egyptian educational system like? In a way, we must term it a phenomenal success. The children learned very well exactly what they were taught:

" 'Then did Pharaoh call for the wise men and the sorcerers.' Then did Pharaoh begin to mock them and crow like a cock, saying to them, 'so these be the signs of your God! . . . Are you not aware that all kinds of magic are within my province?' He then asked for children to be brought from school and they also performed these wonders . . . It is written, 'Then Pharaoh *also* called.' Why *also?* Because his wife whom he called also did this. . . . Even children of four and five years of age did likewise."

Such was the Egyptian educational system, amazingly successful, yet to what end? Every child was by the age of five intimately acquainted with the rites, incantations etc. of the occult. The Egyptians had developed their pedagogical skills to a very high degree. Children at the ages of four and five could be taught all manner of esoteric things. Yet, what good did it do them? The spontaneity of childhood was soon forced into the discipline of witchcraft. The natural good humor and *joie de vivre* of youth were soon perverted into the dark morbidity of necromancy, and in a moment, the natural magic of childhood was gone, miserably replaced by the black magic of the *hartumin.* In truth, Egypt had no real children, only premature youths and immature adults. Such were the products of the elementary schools, high schools and universities of Egypt.

VI. *YOUR OVENS*—Egyptian Industry

Like Noah, who, as we learned in Chapter II, invented the plow to grant leisure to an overworked mankind, many inventors flourished in ancient Egypt. All manner of machines were produced to make the burden of work lighter upon the human race. But, as in Noah's time, the crucial question became:

what will man do with his new leisure time? Like the modern assembly line worker, Egyptian laborers could take little pleasure in their products. The system did not allow the lowly worker to take pride in what he had produced, so he was constantly looking for fulfillment elsewhere. Finding no solace in work or wholesome culture, even the common Egyptian was thrown into the web of decadence and depravity which was spread over Egypt. The *tzfardeim* which attacked the centers of industry brought home to the Egyptian leaders the unhealthiness of their system and the need for immediate change.

VII. *YOUR KNEADING—TROUGHS*—THE MINISTRY OF ECONOMICS

Just as the leaders of Egypt were so far-removed from the common people, that did not see the need for an excellent health-care system, so did they not care about the economic problems the workers had to face. Depressions, Recessions, and other economic calamities had no meaning for the upper classes who lived in opulence and luxury. The *tzfardeim* who destroyed the Ministry of Economics left behind a legacy of the necessity for a total overhaul of the Egyptian Economic system.

THE LAWMAKERS

Until now, we have discussed the first wave of the Frogs' attack, when they exposed the regulatory agencies of Egyptian government. But we will remember from the Biblical passage that there was a second wave of Frogs:

And the frogs shall come up both upon you, and upon your people and upon all your servants." We know that, whereas the first wave was concerned with the exposure of the folly of the *enforcing* bureaus of government, the second wave attacked the source of the problem, the places where the laws are codified. These places generally include three different branches in most democracies:

UPON YOU—THE EXECUTIVE BRANCH

The Hebrew word *bekha* (upon you) appears here for the only time in the Torah with the letter *heh* at the end. The reason is that the Torah wishes to emphasize here the importance of the wisdom of the national leader, and the Hebrew word for wisdom (*hokhmah*), which is feminine in gender, requires the letter *heh*. The national leader of any country, be he king, president, or prime minister, must radiate a wisdom and dedication that can excite and influence his own aides and advisors. If the national leader does not inspire his counselors to exert their maximum efforts for the good of the populace, it is a sign that *he* has not impressed upon them that he cares either.

AND UPON YOUR PEOPLE—THE LEGISLATURE

What is often called "the people's house," be it named congress, house of commons, *knesset*, or whatever, is the branch of any democracy which most closely reflects the views of the people. Its members are the ones held most closely accountable for their actions and report on the most informal basis to their constituents. This can have both an advantage and a dangerous disadvantage. On the one hand, it seems logical that the members of the legislature cannot become as far removed from their constituents as the members of the other two branches of government. There is more of a chance that they will remain "human." Yet, on the other hand, there is also the very great danger that they will be afraid to provide moral and courageous leadership. They will cower before the spectre of popular opinion and will vote not their consciences, but what they think will please the greatest amount of people and offend the least. The *tzfardeim* which attacked the Egyptian "house of the people" showed clearly what kind of "people" and what kind of a "house" it was.

ALL YOUR SERVANTS—THE JUDICIARY

In the end, it is the courts which make the law. A statute can be initiated by the executive branch and enacted by the legislature but it is the interpretation given to it by the court which defines its ultimate character. Undoubtedly, some of the laws of Sodom and Gommoroh looked quite benevolent on the books; it was the way they were carried out that gave those ignominous cities the reputations they deserved. The judge who finds legal hoopholes to free robbers and murderers, but exercises *strictum jus* upon victims of hoaxes and perfidy, is the true villain of a civilization. The *tzfardeim* entered the buildings of Egyptian 'justice' in droves and dispelled any ideas the Egyptian people might have had about the dedication to justice of their "guardians of the law."

THE FROGS AND HISTORY

Of all the Plagues, the only one which is recorded to have had any effects later in history is the Plague of Frogs:

"Tudos, a Roman Jew, expounded: 'From where did Hananiah, Mishael and Azariah obtain proof that they should present themselves to the fire for sanctification of the Lord's name? They drew a *kal v'homer* from the frogs in Egypt to themselves. Since the frogs, who are not commanded to sanctify the name of the Lord, threw themselves . . . into the hot ovens, should we not of course sanctify the name of the Lord by throwing ourselves into the fire! (Pesahim 53b)."

This passage in the Talmud has, of course, elicited much controversy among the commentators and Responsa writers.

Yet, the fact of the matter remains that it was the Frogs who inspired Hananiah, Mishael and Azariah to cause a great *Kiddush Hashem* by flinging themselves into the hot ovens. What was their reward? We find that it came in two stages, immediate and long range:

" 'And the Frogs died from the houses.' Those who were in the *houses* died. But those who had entered the ovens and sanctified the name of the Lord, were rewarded with life."

THE SONG OF THE FROGS

The long-range, indeed eternal, reward of the frogs may be found in a Midrash known as *Perek Shirah*. This Midrash records the song of praise each of God's creatures sings to Him as the object of its creation. Generally, these songs are reflected by verses from the Torah which speak of that particular creature. Thus, for instance, the heavens sing:

"The heavens declare the glory of God, and the firmament shows His handiwork." (Psalms 19:2). And the sun says:

"The sun and the moon stand still in their habitation; as the light of Your arrows they go, At the shining of Your glittering spear" (Habakkuk 3:11). Each creature utters a praise for the Lord which fits its own personal place in creation. What does the frog sing?

"Blessed is the name of His Kingdom Forever and ever." This verse, which even the Jewish people can only declare aloud on the Holy Day of Yom Kippur, is granted to the frogs for all time because they proclaimed the Almighty's Kingdom throughout the world by fulfilling His bidding. Each frog had the choice to go into one of the less dangerous of the "targets," the beds, the courts etc. Since so many of their number chose to sanctify the name of the Lord by giving up their lives, they were forever granted status as the praise-singers of the Lord's kingdom.

THE FROGS AND THE POWER OF FIRE

We can now begin to understand the intricate relationship of the Plague of Frogs with the *Dor Hahaflaga's* sin of misusing the power of fire. On a primary level, the Frogs helped to atone for that great sin by utilizing the fires of Egypt to sanctify the name of the Lord. On an even more profound level, the relationship between the two involves the very origin of fire. Our sages in the Talmud (Pesahim 54a) teach:

"On the evening following the first Sabbath, God granted Adam a special heavenly wisdom and he ground two stones against each other and produced fire."

The Talmud stresses that the discovery of fire was no accident; it was the first creative act by a human being and as such deserves recognition as one of the milestones of history. Just as Egypt, the first *Galut*, becomes a prototype of all future exile, so is the discovery of fire the archetypal creative act. For all future generations, *fire* is the metaphor for the potential creative powers within man. It symbolizes his hopes, aspirations, possibilities, and sadly, the danger of his own self-destruction.

For generations since Adam's momentous discovery, mankind had continued to exhibit "a special heavenly wisdom," but unfortunately he rarely utilized that celestial knowledge to perform the work of his Creator or for the benefit of his fellow man. Things became progressively worse until the Generation of the Exodus when the Egyptians had totally perverted the lofty power of human creativity. The next Plague, *Vermin,* dealt in detail with the various aspects of misused creativity. It was the task of the Frogs to alert the Egyptians to the fact that their *deah* (wisdom) was not of the heavenly type, but was best represented by the *tzfardeah*, the abandonment of wisdom, and the embracement of puerility. The Frogs accomplished this by attacking the symbols of government in Egypt and discrediting them in the eyes of the Egyptian people.

Thus, the way was paved for the Plague of Vermin, when the so-called "wise men" of Egypt were repudiated entirely. Although it was not until the Plague of Locusts that the common people themselves dared to speak up to Pharaoh, nevertheless, it was the Plague of Frogs which began the irreversible process of freeing the "common sense" of the "common" people from the *tzfardeim* of Egypt.

Chapter VI

VERMIN: THE LESSON OF CREATIVITY

It will be recalled that the Plague of Vermin corresponds chronologically to the first sin of the *Dor Hahaflaga*, that is, misuse of the power of the inanimate elements. At first glance, there seems little to connect the two; with a bit further investigation, however, we soon find a deep inner bond between them.

Let us review the few verses in the Torah which speak of the Plague of Vermin:

"And the Lord said unto Moses: 'Say unto Aaron: Stretch out your rod, and smite the dust of the earth, that it may become vermin throughout all the land of Egypt.' And they did so; and Aaron stretched out his hand with his rod, and smote the dust of the earth, and there were vermin throughout all the land of Egypt. And the magicians did so with their secret arts to bring forth vermin, but they could not; and there were vermin upon man, and upon beast. Then the magicians said unto Pharaoh: 'This is the finger of God'; and Pharaoh's heart was hardened, and he hearkened not unto them, as the Lord had spoken."

A careful study of this passage reveals a number of novel aspects of the plague of vermin:

THE FALL OF THE HARTUMIM

a). Until this plague, Pharaoh's sorcerers were called the "*hartumim* of Egypt"; when they are mentioned here, they are referred to simply as "the *hartumim*."

b). The second time the *hartumim* are mentioned, the letter yud is omitted from their name.

c). For the first time, the *hartumim* cannot duplicate any aspect of the plague and are forced to admit that "the finger of God" is at work.

d). After this plague, Pharaoh no longer calls upon his *hartumim* and they are heard from no more in the Torah's account of the events leading up to the Exodus.

THE CREATION OF THE VERMIN

For the first time in the progression of the plagues, something is created. In the Plague of Blood, a *transformation* took place—from water to blood—but no new beings were formed. Similarly, during the Plague of Frogs, no new creatures were brought into being. The Torah uses the words, "the frogs came up," signifying that they gathered together from all parts of the Nile and swarmed over Egypt.

Only in the Plague of Vermin does the Torah use the words, "that it may become vermin." The Hebrew word *l'hotzi*—to bring forth—is used in connection with the vermin, as it was in the act of creation itself: *totzei ha'aretz*—"let the earth bring forth." Clearly, the grand act of creation is not utilized lightly. What is its special relevance to the Plague of Vermin? In addition, a study of the Midrash reveals even more interesting aspects of the Plague of Vermin:

The sage Rabbi Eliezer teaches, "at the moment that the Plague of Vermin reached the bodies of the Egyptians, the yoke of labor and slavery was lifted from our ancestors." The obvious question is, why did physical bondage end at this particular point in the amercement of Egypt.

THE EARTH RETALIATES

In order to properly interpret all of the phenomena mentioned above, we must examine more carefully both the mechanics of the Plague of Vermin and the symbolic meaning of brickwork.

The Torah makes clear that the Plague of Vermin was accomplished through striking the earth: "and Aaron stretched out his hand with his rod, and smote the dust of the earth." For this reason, Aaron performed the task, not Moses, who had once benefited from the earth (Exodus 2:12).

Brick-work, also, is intimately related with the use of the earth for man's purposes. In ancient Egypt, limestone, sandstone, red granite or the black mud of the Nile itself mixed with sand, was used to fashion bricks. We thus begin to see the profound relationship between the *Dor Hahaflaga's* battle cry, "let us make bricks," the Plague of Vermin and the cessation of Jewish brick-making. The relationship is a triangular one:

let us make
bricks out of
the earth

vermin from
the earth

the Jews ceased
to make bricks

The Plague of Vermin taught both the Jews and the Egyptians an invaluable lesson in progressing from the mistakes and sins of the *Dor Hahaflaga.*

THE LESSON FOR THE EGYPTIANS

In demonstrating to the Egyptian "wise men" the impotence of their powers against the omnipotence of the Lord, God specifically chose the two media of earth and vermin. Their interchangability indicated two basic factors in a Divinely structured universe:

1). The powers of uncleanliness and iniquity have no real dominion over any force in the world, even over the tiniest and most helpless of creatures. Indeed, their impotence over the vermin merely proves that any seeming powers they may exhibit over larger creatures are displayed by the grace of God and only when it serves His purposes.

2). The rendering of the inanimate earth into crawling, living creatures demonstrated the unpredictability of even the most sure-seeming "scientific principles," and indicated that they, too, were established by the Creator, to be continued or suspended at His will.

To accentuate this lesson of the limited nature of even the most advanced technological knowledge, the Lord allowed the wise men to grow complacent in their apparent imitation of the first two plagues. The shock of helplessness at the Plague of Vermin was made all the greater by the magicians' ostensible success in duplicating the first two plagues. Enamoured of their great power over the elements, hubristic in their arrogance,

Pharaoh's advisors needed an unimpeachable lesson in Divine Providence and God's absolute sovereignty in the world. This was appropriately accomplished through the Plague of Vermin. Not only did the magicians have no power over their own Egyptian earth, but the very scientific laws they had accepted as sacred had turned upon them.

Thus it was during the Plague of Vermin that the *hartumim*—the representatives of Egypt's proudest scientific achievements—completely lost their effectiveness and credibility. First their title was diminished, then their very name was shortened and eventually they disappeared entirely from the scene.

Once the credibility of the sorcerers and wise men was destroyed forever, the road was opened for lessons in the proper reorientation toward the moral and ethical life. Once man rejected his own omnipotence and accepted Divine Providence, he could make tremendous spiritual progress. The new perceptions gained with the Plague of Vermin must be viewed, therefore, not only as an important milestone in the progression of the ten plagues, but as a turning point in the entire history of mankind.

THE LESSON FOR THE JEWS

For the Jewish people, the Plague of Vermin provided a vital stepping stone on the road to becoming the "nation of priests" which God had destined them to be.

The brick has a symbolic value, in addition to its intrinsic meaning. As a universal metaphor, the brick symbolizes any unit of combination or construction. Thus, the "building blocks" of a table are bricks of wood, of an automobile, of steel etc. Once creation *ex nihilo* has been completed, all of man's endeavors depend upon his use of what we may term "brick technology." What to combine with what, in what proportions and to what end, are the basic questions of modern industry.

As the "bricks" become ever more complicated and the products more esoteric, man slowly begins to forget the "Prime Cause" and the Source of all the elements. He builds his towers so high and becomes so involved in the formation of new and better bricks that he forgets his roots in the earth: "till you return unto the ground; for out of it you were taken; for dust are you and unto dust shall you return" (Genesis 3:19).

To teach man the ultimate lesson of proper and improper creativity, God brought the Plague of Vermin upon Egypt. The Egyptians, who had misused their great stores of knowledge, who had perverted their founts of wisdom, had their own ground turned against them. Yet, man can turn the body which comes from the earth, into an indomitable tool for Divine service, if he only so desires. He can utilize the formerly inanimate clay of his body to imitate his Creator's creativity, if he retains the holiness his Creator granted him.

"With the Plague of Vermin, our ancestors ceased to labor with bricks." This succinct Midrashic statement tells a great spiritual story. As the yoke of physical bondage was lifted, the spiritual light generated by physical freedom bathed the former aspirations of the Jews in the harsh luminescence of truth.

The nation that would soon receive the Torah had much to learn about the tremendous power *and* misuse of human creativity. "Come, let us make bricks," the *Dor Hahaflaga* said. The leaders of that pivotal era had tapped vast reservoirs of human creativity and had proceeded to exploit them without reflection or moral purpose.

The *hartumim* of Egypt also thought they had the power and the *right* to command the elements to do their evil bidding. The Plague of Vermin came to purge the machinations of the alchemists, scientists, sorcerers and magicians. The Jews, now attuned to the didactic messages of the plagues, began to see the folly of their earlier infatuation with Egyptian technology and pseudo-science. With the noble instincts inherited from the Patriarchs reemerging, they immediately changed their goals and aspirations.

No longer were their minds focused on the acquisition of temporary luxuries or the building of useless edifices, but they began to turn their hearts and minds to higher and more noble purposes.

It was only with the demise of the hero-worship accorded the Egyptian "scientists" that the Jews could begin to reclaim their own ancient heritage. It was only with the miraculous creation of new beings that the possibilities for true creativity could be manifested. With the completion of the lesson of human creativity and power, the Egyptians and the Jews were ready for the lesson in the proper limits of that power, the plague of the Wild Animals.

Chapter VII

WILD ANIMALS: THE LESSON OF LIMITATIONS

A LIFE OF ROBBERY

The plague of Wild Animals began the atonement for the sins of the Generation of the Flood. The first and most serious of these was the total pre-occupation with robbery of every sort. Our sages teach that robbery was so much a way of life that they stole almost out of force of habit, without any monetary gain. Someone would steal something one day and the next day his neighbor would steal it back. This pattern grew into an endless cycle, which, as it grew worse, could only lead to ultimate destruction.

Let us attempt to pinpoint exactly the factors within man's nature which cause him to take that which belongs to another. It is clear from the picture our sages paint of the antediluvian era that simple greed and covetousness are not the sole causes of robbery. Noah's invention of the plow had given mankind unprecedented leisure and well-being. Freed from the back-breaking toil of hand-tilling the earth, the Generation of the Flood sought new ways to relieve the sudden boredom and "pass the time." Indeed, our sages (Sanhedrin 108a) teach that the Generation of the Flood enjoyed the ideal conditions for the perfect life:

THE PERFECT LIFE

"The Generation of the Flood waxed haughty only because of the good which the Holy One, blessed is He, lavished upon them. Behold, what is written of them? 'Their houses are safe from fear, neither is the rod of God upon them.' It is also written, 'their bull genders and foils not; their cow calves and casts not her calf.' Further, 'they send forth little ones like a flock and their children dance.' Further, 'They take the timbrel and the harp and rejoice at the sound of the organ.' It is also written, 'they spend their days in prosperity and their years in pleasures'; And it is also written, 'and in a moment go down to the grave.' (i.e., they do not suffer before death). And 'tis that which caused them to say to God, 'depart from us; for we desire not the knowledge of your ways. What is the Almighty that we should serve Him? And what profit should we have, if we pray unto him?' They said thus: Do we need Him for anything but the drop of rain? We have rivers and wells to supply our wants."

We may irrefutably conclude from the words of the Midrash that the antediluvian age was not an impoverished era, when a man stole "to satisfy his soul when he is hungry" (Proverbs 6:30). Rather, the Generation of the Flood flourished in a utopian age, when all the blessings of heaven were granted to mankind. How, then, do we explain the widespread epidemic of thievery?

MAN: THE DIALECTICAL CREATURE

The answer lies within the very essence of the nature of man. Man has been described accurately as "a dialectical creature," in constant conflict with the various aspects of his own psyche. On the one hand, man is composed of an infinite and eternal soul. On the other, he is a woefully finite creature, constantly deteriorating physically until he reaches the grave.

This dual nature in man's disposition leads to various conflicts in the proper orientation of his aspirations. His eternal soul yearns for the infinite, the unreachable, for the perfection unattainable in its earthly state. The finite body seeks the ephemeral pleasures and joys that will nourish, sustain and perhaps even extend its own tenuous existence.

Yet, there come times when the aspirations become confused. The body yearns for that which is legally or morally beyond its reach and the soul becomes so polluted it can no longer restrain its corporeal partner. It is at that point, when the body loses sight of its proper limitations and achieves mastery over the ethereal soul, that the quintessential act of stealing occurs.

THE FIRST ROBBERY

The very first act of stealing occurred when man was indeed lord and master over all that was. God had placed only one limitation on Adam: "Of every tree of the garden you may freely eat, but of the tree of the knowledge of good and evil you shall not eat." Nevertheless, man ate; thus, the Midrash refers to Adam as a thief. In taking from the fruit of the forbidden tree, Adam *overextended* his sovereignty—and thus committed the act of robbery. We therefore may conclude that thievery does not necessarily derive from a feeling of deficiency, but from the desire to overextend the sphere of one's influence or domain.

THE UNIVERSAL LIMITATIONS

The concept of fixed and definitive "spheres of influence" is one of the primary universal functions. The orbits of planets, the crucial distances between stars and even galaxies, the territorial integrity of nations, the private ownership rights of individuals, all depend for their continued existence upon the law of limitations.

Scientists in every field have proven mathematically the disastrous results that could occur if certain functions broke free of their sphere of limitations. If the moon circled any closer to the earth, catastrophic tidal waves would result; if it suddenly widened its orbit, the lack of tide would be equally calamitous. If the earth orbited closer to the sun, our entire planet might be turned into a flaming inferno; if we traveled further out into space, we might become a frigid Plutonian iceberg. The importance of rigorous limitations is thus clearly established on the cosmic level.

On our own everyday earthly level, limitation is equally important. To be sure, man's place in nature's hierarchy is a majestic one. But man must still "know his place," in relation to the world of nature, and especially in relation to his fellow man.

In embracing lives of thievery, the Generation of the Flood sought to break all boundries set upon them. They attempted to throw off all limitations placed upon them by their Creator and be "free" of any restricting rules or laws. As we shall discuss in the chapter on the Plague of Boils, their sexual depravity also "knew no bounds." To them, everything was permissable; nothing was limited.

In ancient Egypt, the rebellion against limitations of any kind reached a new moral low. Living in opulent decadence, the upper-class Egyptians could get almost anything their hearts desired. Yet, to the sorcerer-scientists, this was not enough. They wished to harness not only what man had historically utilized for his well-being, but forces traditionally denied to man.

THE INTEGRITY OF THE ANIMAL WORLD

The words of the Midrash are illuminating: "Why did the Lord bring the plague of Wild Animals upon them? Because when the Jews ceased to work with bricks, the Egyptians sent the Jews out to the deserts saying, 'let them capture for us bears, lions, and leopards for use in our wars.' Thereupon Moses went and declared before the Lord, 'Can then a human being prevail against a lion?' To this, the Lord answered, 'fear not; go to the fields and call to the fierce beasts and say to them: Thus says the Lord, for this I have prepared you from the day I created you. They shall all come upon Egypt in mixed droves and destroy the country.' Moses went and did exactly as he had been told, and so it was as the Lord had spoken."

We learn from this Midrash that the Egyptian wise men attempted to overextend themselves in their relationship with the animal world. For generations before the Egyptian civilization, it had been known and accepted that the animal kingdom was separated into two distinct parts: those who could be domesticated and those whom man had to allow to remain wild.

In general, mankind had accepted this dichotomy and had not attempted to interfere with the obvious pattern of God's plan in nature. Only the Egyptians took upon themselves the sinister task of intervening in God's blueprint for the world.

With their egos dangerously inflated by continuous scientific and technological successes, the Egyptian *hartumim* could not bear the thought that a part of nature was not under their control. Yet, recognizing the danger inherent in their self-appointed task, they sought to force the Jewish nation to perform the task for them.

To understand how befitting was the Plague of Wild Animals as a punishment for this sin, we may apply a saying of the Talmud:

"Whoever sets his eyes on that which is not his, is not granted what he seeks and what he possesses is taken from him" (Sotah 9b).

The Egyptians, who had sought to pervert nature and utilize wild animals in immoral and unnatural ways, had their machinations turned against them. Not only could they not control the animals of the wild, but the beasts deliberately left their natural habitats to attack their would-be tamers.

The Plague of Wild Animals, which came to teach the antidote to the atmosphere of robbery and rapacity of the *Dor Hamabul* did its job well. As the Jewish nation watched the wild animals emerging from the deserts to attack the Egyptian powerful-elite, they absorbed the lesson of the importance of human limitations. Millenia before any formal environmentalist or ecologist, the fourth plague taught the lesson of respecting nature's precarious equilibrium.

On an even more profound level, the Plague of Wild Animals reflects a historical turning point in man's relationship with nature. We know that the animal world was created to serve man. For this reason, man was created last, so that all of creation would be waiting to serve him. We know, also, that all the animals passed meekly under Adam's watchful eye as he assigned them their proper names. Surely, no animal dreamed of attacking Adam then. Yet today, no sane person would dream of approaching a fierce wild animal in friendship, let alone superiority. What happened in the interim?

CAIN AND THE ANIMALS

For the answer, we must go back to the story of Cain. After he had killed his brother and had been cursed by the Lord for his terrible sin, he complained that "whosoever finds me will slay me." Thereupon, God granted him a sign "lest any finding him should smite him."

Now, who was Cain so afraid would set out to kill him? The only other human beings on earth were his parents, and despite his wicked act, it seems unlikely that Cain was afraid they would slay him. The answer is, of course, that for the first time, a human being began to fear that which was below him, the animal world. Until Cain's grisly act, the condition of "and the fear of you and the dread of you shall be upon every beast of the earth" (Genesis 9:2) reigned supreme. Yet, this natural dominion of man over beast was predicated upon one caveat: man's adherence to Divine law.

Once Cain had broken God's law, he was no longer a step above the animals, and he began to fear their natural physical superiority. Our sages (Sanhedrin 38b) deduced this relationship between man and the animal world from a passage in Psalms:

"A wild beast has no dominion over man unless he appears to it as a brute, for it is written, 'Men are overruled when they appear as beasts' (Psalms 49:13)."

THE FORESTS ARE CLOSED TO MAN

As long as mankind had remained morally and spiritually above the bestial level, it retained total dominion over nature.

Yet, once man descended from the pedestal upon which his Creator had placed him, the entire course of civilization had to be changed.

Whereas once, mankind could roam nature freely, exploring the open and inviting garden full of benevolent adventure and readily available treasures, once man lost his grandeur, the world of nature became for him a deadly trap full of potential attackers and ubiquitous danger.

Given his new fear of the outdoors, there was only one thing man could do. He began to erect barriers and walls separating himself from the once cherished world of nature. Thus, it is no coincidence that Cain was the first to build a city. If he had not lost the respect of nature, there would be no need to sever himself from it. If he had remained superior to the animals, he could have dwelt among them and had them for willing and eager servants. But of course, Cain did not realize the full significance of his sin and its retribution, and he proceeded to pervert even the enforced urban living, as we have already discussed.

THE TRESPASSERS

Thus mankind accepted, albeit with a heavy heart, the new boundaries which had been thrust upon it, until the advent of the Egyptian civilization. Pharaoh and his magicians would not accept the limitations mankind had lived with since the age of Cain. Yet, neither did they attempt to return to the state of Pre-Fall innocence that had allowed man to retain dominion over beast. In seeking to subjugate the wild animals, over whom they could claim absolutely no moral or spiritual superiority, the Egyptians committed the sin of robbery in its most elemental form. They trespassed upon the very boundaries which God had set for errant mankind.

What made Pharaoh think he could succeed where generations had not even dared to try or had failed miserably? The Egyptians were blinded by their own so-called technological achievements. Having harnessed to an almost unparalleled degree, the powers of human creativity, they thought that they could control animal nature as well. And there, they erred grievously. For Pharaoh's magicians did not take into consideration the extraordinary sensitivity God had instilled into the animal world. The wild and fierce animals which attacked Egypt felt instinctively the moral depravity of their would-be tamers and thus were rendered all the more uncontrollable. The Egyptians did indeed appear to the animal world "as brutes" and there was therefore nothing to withhold them from their attack.

MOSES VISITS THE NILE

We can now understand a hitherto inexplicable aspect of the plague of Wild Animals. Before the onset of the plague, God instructed Moses to "Rise up early in the morning and stand before Pharaoh; lo, he comes forth to the water." The only other occasion when Moses is commanded to seek out Pharaoh at the water is before the Plague of Blood.

In the case of the Plague of Blood, we understand the relationship because the plague will strike the water itself. But why before the Plague of Wild Animals? Furthermore, in the Plague of Wild Animals, Moses must antecede Pharaoh and stop him *before* he reaches the Nile. What is the significance of this command?

We will recall from our discussion of the Plagues of Blood and Frogs that the Nile symbolized the accumulated wisdom and technology of the Egyptian civilization. In the Plague of Wild Animals, it was graphically illustrated to the Egyptians that not only had they overstepped natural bounds in their

Faustian use of knowledge, but that their wisdom had not raised their stature at all in the grand scheme of nature.

While the Plagues of Blood and Frogs emphasized the corrupt nature of Egyptian "wisdom," the Plague of Wild Animals demonstrated that the Egyptians had not yet even begun to rise above the animals. Their wisdom was nothing, and even their humanity went unrecognized and unacknowledged by the beasts of the field. Moses encountered Pharaoh even *before* he went down to the Nile to show him that *this* plague would teach him that he had not yet even arrived at the beginnings of wisdom, let alone mastery and control of the animal and mineral world.

We are now ready to see how the next logical step was the removal of the Egyptian domination over even the domestic animals, whose significance will be discussed in the following chapter.

Chapter VIII

PESTILENCE: THE LESSON OF POWER

THE RETURNING PLAGUE

While each of the Plagues atoned for a specific sin of mankind, *Dever* (Pestilence) provided a generic recurring lesson on the underlying cause of all mankind's ills, the lust for power. We know that the message of *Dever* was the most crucial one of all because it was the only one which was repeated. The Midrash makes the extraordinary statement that "every Plague visited upon the Egyptians was accompanied by a Pestilence." This revelation should alert us to the fact that after each individual lesson concerning a particular human failing, it was necessary for the Egyptians to be reminded of the dangers of excessive power.

This matter can best be explained with an analogy. On the Sabbath preceding the New Moon, a special prayer for the coming month is recited. In this prayer, we ask for many blessings:

"O grant us a long life, a life of peace and well-being, a life of blessing and sustenance, a life of physical health, a life with *fear of Heaven* and dread of sin, a life free from shame and disgrace, a life of wealth and honor, a life marked by our love for Torah and our *fear of Heaven.*"

ANOTHER DOSE OF THE FEAR OF HEAVEN

Many of the commentators upon the Siddur have noted that for some reason, the prayer for the character trait of "fear of Heaven" is repeated. Rabbi Israel Meir Kagan, the gentle sage known as the *Hafets Hayyim*, notes succinctly: "Once a person has achieved 'wealth and honor,' he must pray again for 'fear of Heaven.' "

The *Hafets Hayyim's* elucidation of the prayer for a happy month helps us to understand the need for repetition of the Plague of Pestilence. We know that the Egyptians lived in luxurious decadence. Each of the Plagues removed a little bit of that luxury, but left the Egyptians enough to give them an opportunity to repent on their own. When the Egyptians went back to the tranquility of their mansions, with their servants and concubines, riches and abundance, they often became once again intoxicated with their power and drunk with the reach of their influence.

Only the Plague of Pestilence was able to wean them of this inebriation and return them to the sobriety of the truth. The Plague of Wild Animals had permanently cured the Egyptians of their experimentation with the freedom of wild animals. Yet, the large stables of domestic animals which the Egyptians kept were constant reminders of the power which they *did* hold over at least one segment of nature, the domestic animals. Until they could be convinced that even this dominion could be changed, the Egyptian people would not be able to be brought out of their moral dungeons.

A MIGHTY HAND

This interpretation of the significance of the Plague of Pestilence will enable us to understand two difficult aspects of the Plague. We will recall from the chapter on the Plague of Blood, that only two Plagues are referred to in the Hagadah as being more miraculous than the others:

"And the Lord brought us forth out of Egypt with a mighty hand, and with an outstretched arm, and with great awe, and with signs and with wonders" (Deuteronomy 26:8). *With a mighty hand* refers to Pestilence. . . . *With an outstretched arm* refers to the sword. . . . *With great awe* refers to Divine Revelation. . . . *With signs* refers to the staff. . . . *With wonders* . . . refers to the Plague of Blood."

We have already discussed the references to the Plague of Blood in Chapter IV. Now that we understand the necessity for the repetition of the Plague of Pestilence with each of the Plagues, we can appreciate why only *Dever* is referred to as having been delivered "with a mighty hand." Not only was the Plague of Pestilence repeated, but it continued to chip away at the Egyptian addiction to power until it had completely conquered this dangerous craving.

THE MERCIFUL PLAGUE

Another fascinating aspect of the Plague of Pestilence concerns a unique word the Torah uses in this connection:

"Behold the hand of the Lord *hoya* (is) upon your cattle."

As the Midrash points out, "there is nothing like it in all of Scripture." The word *hoya* does not occur anywhere else in the Torah and presents a mystery to us in explicating its meaning.

Even at first glance, we perceive two aspects of the word *hoya*, even if we cannot compare its use with any other passage. It seems to combine the words *haya*, *hoveh* and *yihyeh* (was, is, will be) and presents, as Rabbi Samson Raphael Hirsch puts it, a kind of timelessness, "something that has no past and no future."

The other aspect, which seems even more abstruse, is its similarity to the Divine name, of four letters. We know that of the two most often used Divine designations in the Torah, the name *Elokim* refers to God's attribute of justice and the name of four letters refers to the Lord as The Merciful One. It is therefore greatly perplexing to find here a form of the Divine name signifying mercy. For surely, as we have learned, the Plague of Pestilence, with its repetitions, was one of the sternest of the Plagues. Where does the Divine trait of mercy fit in?

The answer also lies in the fact of *Dever's* ubiquity among the Plagues. In case the Egyptians should think that all these disasters that were happening to them were unrelated disjointed phenomena, the Pestilence was always there to remind them they were being taught a meticulously planned lesson and they could "graduate" immediately, if they only repented for their evil ways.

The Pestilence reminded Egypt that, far from simply being punished, God was taking great pains to uplift them out of their moral and spiritual prisons. Just as the Jewish people actually did rise through the forty-nine levels of uncleanliness, so could the Egyptians have progressed spiritually and morally during the Plagues. The Plague of Pestilence reminded the Egyptians that it was God's attribute of Mercy which was conducting the Plagues, and not Justice, for were it simply Justice, they would have been wiped out long ago.

Yet, as the Egyptians returned home from each Plague, with their lust for power intact, with the thought that they could still control their slaves and servants, if not their dying animals, God brought yet another Plague upon them. This time, they would have to learn that not only had they no power over the outside world; they did not even control their own bodies. This was the Plague of Boils.

Chapter IX

BOILS: THE LESSON OF REGULATION

The middle three plagues—known by the acronym *ADASH*—were all concerned with the various aspects of domination or ownership. Wild Animals corresponded to the *Dor Hamabul's* preoccupation with robbery, and Pestilence atoned for mankind's oft perverted relationship with the outside world. Yet, the Plague of *Sh'hin*—Boils—corresponded to the most subtle relationship of all; man's relationship with his own body.

We know that both the *Dor Hamabul* and the ancient Egyptians were trapped in the deepest webs of sexual depravity and perversion. Of the *Dor Hamabul* it is said that they even caused the animal world to grow perverse. The Egyptians were well known to be the most sexually immoral nation on earth. Since they abused the bodies which were given to man for a noble and high purpose, their bodies were plagued by ugly sores and warts which called attention to the abuse with which they had treated their own bodies.

Such is the surface relevance of the Plague of *Sh'hin* to the sin of carnality. But there is an even deeper relationship. To penetrate its meaning, we must investigate the dialectical nature of the Jewish attitude toward the body and its needs.

SUICIDE IS MURDER ALSO

Man has both a positive and negative relationship with his physical self. On the one hand, the Code of Jewish Laws decrees: "man does not have ownership rights over his body." He is not free to do damage to himself, or to allow others to do damage to him. If he confesses to a crime for which there are no witnesses, the court does not punish him, for he has no right to bring pain or destruction upon himself. He may not commit suicide, for his body is his only upon loan to be treated gently, with respect and the maximum of care.

Thus, man's treatment of his body is severely limited by the Torah. Man may at times wish to do that which is injurious to his body but the laws of the Torah stop him and say, "thief, touch not that which is not your own." Man may feel that his body is the last bastion of the personal, the one final object of total ownership—and the Torah reminds him that he is mistaken. On the contrary, the body is that which he owns least, for it houses the Godly soul and becomes totally subjugated to its ethereal wishes and needs.

THE BODY IS AN IMPORTANT PARTNER

Yet, while in Jewish tradition, the body is subject to the requirements of its loftier partner, the soul, and cannot be endangered because of the honored guest it carries, its own needs are not to be ignored. Unlike religions which advocate celibacy and asceticism, Judaism recognizes the body as an irreplacable partner in Divine service. Thus, the first commandment in the Torah is "be fruitful and multiply." Far from Original Sin, the Torah recognizes the sexual union of man and wife as the pinnacle of creative mankind's accomplishment. On the Sabbath and Holidays, we are enjoined to partake of and *enjoy* meat, wine and other delicacies.

How does the Torah reconcile for us the lack of freedom over our physical selves, yet the allowance—indeed requirement—to participate in the bodily pleasures? The answer lies in the Torah's concept of *regulation.* Left alone, without discipline or control, the physical passions can wreak havoc on our bodies, destroying on the way the delicate equilibria of our constitutions. Utilized properly, guided and regulated, they can make our lives pleasant and enjoyable, giving us the physical strength and vitality to happily fulfill the Divine will.

The Torah solves for man his eternal conflict between body and soul, between physical and spiritual, by acknowledging the corporeal side of human nature and utilizing it within a spiritual context. Thus, the Hebrew word for marriage is *Kiddushin*—holiness—signifying the lofty ideal of married life recognized and expected by the Torah. Rather than ignoring the physical aspects of life, the Talmud discusses in great detail the *laws* of bodily necessities, sexual conduct, menstruating etc. Since *istakel b'oreita ubara alma*—"G-d utilized the Torah as the blueprint for the world"—there is no natural phenomenon, no aspect of human life, no element of nature, that is not discussed in the Torah.

The Jewish attitude toward the place of the physical aspects of man in his spiritual life, may be described in Maimonidean terms as the *derekh haemtzait,* or "the golden mean." On the one hand are the religions which, in their inability to deal with physical man, choose to ignore the corporeal side of life altogether. They ban their holy men from family life (whereas in Jewish law, a childless person or bachelor was not allowed to be a judge) and preach an ascetism and hermitic life which ignores all aspects of human nature with which it cannot deal.

At the other extreme are the cults and sects which glorify the body and whose ceremonies degenerate into all manner of immoral sybaritic rites. Egypt represented the nadir of such practices, being known even among the barbaric nations as orgiastic and totally uninhibited. The Torah issues a special admonition against following in the ways of our ancient task-masters:

"After the doings of the land of Egypt, wherein you dwelt, shall you not do" (Leviticus 18:3).

We may recall that the Egyptians are not blamed by the Torah for deliberately *luring* the Jews into immorality. Rather, by the time of the Jewish sojourn in Egypt, it was the *essence* of the Egyptians that they were inexorably entwined in carnality. The Maharal of Prague uses this illuminating fact to explain why God chose Egypt as the place of the first and archetypal *Galut*:

WHY EGYPT?

"When the Lord sought to fulfill the decree He had made to Abraham, 'Your seed shall be a stranger in a land that is not theirs, and shall serve them; and they shall afflict them four hundred years,' this had to be accomplished amongst a nation which was the direct antithesis of Israel. In what way are Israel and the Egyptians opposites? The ancient Egyptians were totally entangled in carnality and their very natures were attuned only to immoral and abominable practices. However, it is the essence of Israel that they are holy and piously abstain from all manner of immorality. Therefore, it was logical that Israel begin their period of slavery among a people who were their exact opposites in temperament, nature, and indeed geneology."

With this scintillating commentary of the Maharal in mind, we may begin to understand a unique aspect of the Plague of *Sh'hin*. The Midrash notes that the plagues may be delineated according to who actually brought about the final phenomenon:

"Three of the plagues came through the agency of Aaron, three through Moses, three through God and one through the united efforts of all three. Blood, Frogs, and Vermin . . . were through Aaron; Hail, Locusts and Darkness through Moses . . . Wild Animals, Pestilence and the Plague of the firstborn through God, and the Boils—through all of them."

In another place, the Midrash is even more specific about the singularity of the Plague of Boils:

"A great miracle happened in connection with the Boils. Both Moses and Aaron took handfuls, but the hands of Moses held not only his handful, but also that of Aaron, a proof that the lesser can contain the greater. This plague was brought about by God, Moses and Aaron, because Moses and Aaron took handfuls of soot and threw them heavenwards, and God converted them into boils while still on high, and these descended upon them."

The inevitable question which presents itself upon studying these two Midrashim is: Why was the Plague of Boils singled out to necessitate the action of God, Moses and Aaron? Surely, the Plague of the Firstborn was more destructive, the Plague of Vermin involved elements of creation etc. In light of all we have learned about the significance of the Plague of Boils, we can appreciate the need for the total participation of all those involved in the bringing of the Plagues.

THE TURNING POINT IN THE PLAGUES

For in a sense, the Plague of Boils, at the center of the Plagues, summed up and epitomized the Egyptian character. It mirrored on the surface the inner ugliness and putrescence that had taken over the personality of Egypt. It is no coincidence that the Plague of Boils was Pharaoh's last chance to repudiate Egypt's heritage of moral and spiritual shame. Once Pharaoh refused to heed the warning of the Plague of Boils, his Free Will was taken away and he was no longer allowed to relent. The Plague of Boils was the most obvious of all the Plagues in its plea to the Egyptians to desist from their immoral manner of living and return to the requirements of the seven Noahide laws. Since Pharaoh and his people showed no sign of relenting, their power of self-determination was removed so that the remaining four Plagues could be brought upon them as a lesson for Israel and to the rest of the world.

Since the Plague of *Sh'hin* corresponded to that which was most uniquely Egyptian, the total immersion in carnality, *Sh'hin* became the focal point of all the Plagues. Just as all aspects of Egyptian life were mirrored in the Plague of *Sh'hin*, for their erotomania was all-pervasive, so was it necessary for the Lord, Moses and Aaron to unite in the performance of the Plague itself. The paramount importance of this Plague could be conveyed only by singling it out as the only Plague to involve God, Moses and Aaron.

The Plague of *Sh'hin* gave every Egyptian the opportunity to witness and *feel* for himself the contempt in which the body can be held. Virtually every Egyptian, his body rendered repulsive because of the Plague, cursed the physical side of human nature and longed for release from the loathesome body. To make the point of the Plague even more clear, the boils began with the *hartumim*—the worst offenders of all—and then radiated outward to all of Egypt. As in the Plague of Frogs, when the functions of government were struck to show where the fault lay, the boils began on the bodies of the "wise men" to indicate the source of the trouble.

The message from God was clear and distinct: either you will change your odious ways or you will be destroyed. Since they did not heed this plague which sought to change their very essence, the fate of the Egyptians was sealed and they could no longer change their *minds*, for they had not changed their *hearts*.

The insights we have gained into the Plague of Boils will enable us to understand a widely misinterpreted story in the Talmud:

TO BE OR NOT TO BE

"For two and a half years, the school of Shammai and Hillel were debating: one side claimed that it would have been better for man never to have been created and the other side asserted it was better for man to have been created. Finally, they decided that actually, it would have been better if man had never been created at all, but now that he has in fact been created, let him very carefully examine all of his actions" (Eruvin 13b).

It seems incredible that the great schools of the sages Shammai and Hillel should conclude that a person would be better off never to have been created. The Talmud is replete with references to the great benefaction which the bestowal of life is considered by the Torah. We know that our Torah glorifies life—"and you shall live by them (the laws of the Torah)," and the Talmud adds, "and not die because of them" (Yuma 85b). It is almost inconceivable that the sages of the Torah should conclude that if life were not an unwelcome necessity, we should choose not to have lived at all!

THE DANGER OF GOOD HEALTH

If we examine the Hebrew phrase for the creation of man—*b'riut ha'adam*—we notice that it can also mean "the health of the person." Thus, the long dispute between *Beit Shammai* and *Beit Hillel* may have centered around the question: is a life of good health desirable or perhaps is it better for a person to live a life of physical misery and anguish?

This question is quite a different one from that of being created or non-existent and we can well understand how the arguments raged back and forth. Those who thought a life of good health was ideal considered the many wasted hours spent by the sick, unable to properly serve the Lord. They considered carefully the necessity for peace of mind when contemplating the Divine and celestial matters. They thought of the many commandments which cannot be fulfilled by the sick and disabled. And so they concluded that good health was a prerequisite for acting as a servant of the Lord.

Yet, those on the other side argued with equal power. They recalled the extraordinary good health of the *Dor Hamabul* and the destruction to which it led them. They thought of the medical miracles of ancient Egypt and the calamity which it brought upon them. And, above all, they must have remembered the Plague of Boils, and the lesson of Egyptian devotion to the body and the final cataclysm that was brought upon them.

The final decision was perhaps inevitable. It would, in fact, be better for man's spiritual nature if his physical side was neglected and allowed to deteriorate. It might be ideal if man could not be led astray by the desires of his body, because it was a broken and ragged example of the less important side of life.

But, blissfully, that was not how God created the world. Clearly, God wishes us to have our good health and indeed the commandment, "and you shall take care of your souls" includes the proscription to maintain good health standards. How, then, should man avoid the pitfalls of the *Dor Hamabul* and the ancient Egyptians? Upon this point, the schools of Shammai and Hillel were in rare agreement: While keeping the body in the best of health and giving it excellent care, man must scrupulously watch his every action. Good health leads to exhuberance and impetuosity and then, if one is not careful, to thoughtless and irrevocable action.

The Plague of Boils provided an eternal lesson to mankind in the proper contract for the partnership between body and soul. Man can enjoy good health and vitality *if* the soul holds the reins. If the body is allowed to remain unregulated and uncontrolled, boils and other diseases must result to remind man of his true nature and of the eternal soul which he is disregarding.

After the Plague of *Sh'hin,* which defined the essence of the Egyptians and permanently removed any desire for imitating them from the Jewish people, the remaining Plagues could concentrate on the innate needs and problems of mankind. With the Plague of Boils, we are finished with the Egyptians and turn to the inner Man himself.

Chapter X

HAIL: THE LESSON OF LISTENING

ALL MY PLAGUES

One of the verses in the Torah which seems to describe the Plague of Hail, has stimulated a great controversy among the commentators upon the Bible. Immediately after the Plague of Boils and before the Plague of Hail, God tells Moses to warn Pharaoh:

"For this time I will send all My Plagues upon your person, and upon your servants, and upon your people; that you may know that there is none like Me in all the earth." Rashi seems to interpret this passage as referring to the Plague of the First-born. The difficulty in this interpretation can readily be seen upon simply reading the entire chapter. The Plague of the First-born does not come until much later and the warning was delivered just prior to the Plague of Hail. It is also not possible that this warning meant that many more plagues were to come, because the Lord said specifically "this time."

Therefore, some commentators reinterpret Rashi's words as meaning that the text does indeed refer to the Plague of Hail. Yet, this leaves us with almost as great a difficulty: Why is the Plague of Hail referred to as "all my Plagues?" Even if we are willing to accept this designation of the Plague of Hail, a number of other singular aspects of the Plague call themselves to our attention:

THE HEAVENLY CACOPHONY

When Moses told Pharaoh about the coming Plague, he mentioned only "a very heavy hail." Yet, when the actual Plague came, the Torah relates that there was "thunder and hail, and fire ran down the earth." In fact, it is clear that of these three elements, the worst was the thunder, for when Pharaoh begged Moses to have them cease, he said, "Entreat the Lord and let there be enough of these mighty thunderings and hail." When Moses promised to pray for an end to the Plague, he, too, gave priority to the thunder: "As soon as I am out of the city, I will spread forth my hands unto the Lord and the thunders shall cease; neither shall there be any more hail." When God carried out His word and stopped the Plague, the same order was kept: "And the thunders and hail ceased and the rain was not poured upon the earth." Yet, when Pharaoh hardened his heart and did not heed the warning of the Plague, we suddenly find that the order has been reversed: "And when Pharaoh saw that the rain and hail and the thunders were ceased, he sinned yet more, and hardened his heart, he and his servants." Why the sudden change in emphasis to the other aspects of the Plague?

We can best understand the significance of the thunders of *Barad* by investigating *Barad's* place in the overall pattern of the Plagues. We know that *Barad* atoned for Adam's sin of eating from the tree. If we reread the passages where Adam is chastised for his sin, we notice that prominence is given to his original listening to Eve's suggestion that he partake of the fruit: "because you have hearkened unto the voice of your wife and have eaten of the tree . . ." The primary remonstrance concerning Adam's act of *listening* to the entreaties of his wife.

THE MOST DELICATE ORGAN

We can now begin to understand the relationship between Adam's part in the First Sin and the lesson of the Plague of Hail. The awesome sounds of the *Barad* were a warning to mankind to take meticulous care of that most delicate of instruments, the human ear, and to beware of what goes into it and how the information is used. Two of the sayings of our sages are particularly relevant to our discussion:

"Rabbi Levi said: 'six organs serve a man; over three he is master and over three he is not. Over the eye, ear and nose he is not master for he sees what he does not wish to see, and hears and smells what he does not wish to hear and smell. Over the foot, mouth and hand he is master."

Nevertheless, the Talmud (Ketubot 5b) admonishes:

"A person should be very careful not to allow worthless matters to enter his ears, for they are the most delicate and easily destroyed of man's organs."

THE GATEWAY TO THE MIND

In his commentary upon this passage of the Talmud, the Maharal of Prague adds a new dimension to our understanding of the sense of hearing. He explains that the ears are considered the door or gateway to a person. Just as a house would be considered worthless without an entrance, so is the entrance man presents to the world, his sense of hearing. For this reason, although the Talmud (Hagigah 3a) recognizes that a deaf person can often be a great scholar, he is nevertheless exempted by Biblical law from fulfilling the commandments. Also, we might add, for this same reason, if a person causes another to lose his sense of hearing, he must pay him his entire measurable monetary worth, for he has symbolically destroyed him.

We can clearly see, therefore, that the Torah holds the sense of hearing in extremely high esteem and considers it the most sensitive and crucial of man's links with the outside world. If the proper sounds enter man's ears, he is spiritually nourished and strengthened by them. If the wrong words and sounds gain access to his mind, they can do irreparable damage to his soul, psyche and personality and drastic measures may be needed to eliminate their pernicious effects. One such drastic measure was the Plague of Hail.

We can now understand the seemingly contradictory references to the thunders of the Plague of Hail. God intended the Plague of Hail to correspond to and atone for the sin of not listening to *Divine* imperatives and Adam's transgression of listening to his wife and not to the Lord. For this reason, in the warnings to Pharaoh and in the actual manifestation of the Plague, it was the *kolot* (sounds, in the plural) which were the prime factor in the Plague. Pharaoh, too, during the Plague, recognized their importance and wished to have them removed. By granting them the prominence they deserved, he was admitting to Moses and the Lord that he had learned the Lesson of Listening and would henceforth *heed* the word of the Lord. However, Pharaoh was a creature of the moment, whose only thought was for his *immediate* pleasure and comfort. As soon as the cacophony was no longer a present threat and and his ear was not being assaulted by the mounting decibels, Pharaoh obstinately denied their importance. Thus, only when the Plague was over did Pharaoh dare to mention the thunders last:

PHARAOH'S SELECTIVE DEAFNESS

"And when Pharaoh saw that the rain and the hail and the thunders were ceased, he sinned yet more. The Torah even records that Pharaoh *saw* that the Plague was over; he did not *hear* that the sounds had ceased, and they had never really penetrated his inner sense of hearing. Pharaoh's gateway was open only for the Augean aspects of the world, but not for the word of the Lord.

We may now return to the problem of understanding why the Plague of Hail should be considered "all My Plagues." Just as the sense of hearing is considered to be the most important of the five senses, so was the Plague of Hail in certain ways the most important of all. There is no way to learn if one does not *listen.* This listening can be understood symbolically to include other forms of acceptance than that of the audiogenic. Man must concentrate all of his information-gathering faculties to *listen* for what his Creator wishes of him and then to follow up his information with actions.

Although Pharaoh's heart had been permanently hardened after the Plague of Boils, still the Children of Israel continued to learn from the remaining Plagues the lessons they would soon need to receive the Holy Torah. The first of these, was appropriately that of how to listen. This important lesson, which we learned well with the Plague of Hail, resulted in the proudest Jewish statement of all time:

"We will do and we will listen" (Exodus 24:7). By that time, the Jewish faculty of listening was so well developed and attuned, that there was no need for involved explanations, proofs or inducements. The listening immediately resulted in action, so much so, that it seemed as if the action preceded the words themselves. Actually, the Jewish commitment to *listening* to the word of the Lord, made the corresponding action *inevitable.*

For this reason, the Jews at Mount Sinai experienced a once in history phenomenon:

"And all the people *saw the sounds*" (Exodus 20:15). They had developed their sense of *listening* so well that it permeated *all* their senses and there were no longer any neurological boundries between sight and sound. They "saw the sounds."

It is no coincidence that the Plague which followed Hail was that of Locusts, which concerned itself with the sense of sight.

Chapter XI

LOCUSTS: THE LESSON OF SIGHT

THE PARTNER PLAGUES

The Torah makes clear that there was an intricate relationship between the Plagues of Hail and Locusts: Concerning the Plague of Locusts, God warned:

"And they shall cover the face of the earth, that one shall not be able to see the earth; and they shall eat the residue of that which is escaped, which remains unto you from the Hail; and they shall eat every tree which grows for you out of the field." Not one of the other Plagues makes any mention of a prior Plague. Also, there was no connection between any of the plagues, as there was between *Barad* and *Arbeh*, with the suggestion that the second would finish where the first had left off. This proves that Locusts not only followed Hail logically, but was designed to complete its lesson.

This idea becomes clear when we realize that while Hail atoned for Adam's sin of not listening to the Lord, heeding the entreaties of his wife instead, Locusts atoned for Eve's part in the First Sin. We will recall that Eve's sin began with the first fateful viewing of the Forbidden Fruit:

"And when the woman *saw* that the tree was good for food and that it was a *delight to the eyes* . . . she took of the fruit thereof and did eat."

WOMAN SEES AND MAN LISTENS

It is no coincidence that the man's sin is described in terms of listening and the woman's in terms of sight. The sense of hearing is primarily an intellectual function. Most sounds which have an important effect on the human being are converted by the brain into cogent words and thereby register their consequences. The sense of sight is primarily an emotional function appealing instantaneously to the seat of the emotions and eliciting a reaction. Male and female differ in this way also. Although the female may be capable of great intellectual achievement and the male may rise to great heights of emotion, the primary sphere of the female is that of emotion and of the male that of reason.

Because of this general pattern, the Serpent first approached the woman with the temptation to use her power of sight, to look at the Tree with optative eyes and be seduced by the beauty of the tree. The man, on the other hand, had to be won over by the power of speech, by careful and calculated use of words and mental images. The Talmud and Midrash carefully record this monumental conversation between Adam and Eve before he fatefully ate from the fruit.

Since the sense of sight is so closely bound to the sense of sound, the Plagues of *Barad* and *Arbeh* are also closely linked together and Pharaoh is warned that whatever is left over from the Hail will be destroyed by the Locusts. Symbolically, he and the Nation of Israel were being warned about the baleful influences of misuse of the senses. Properly used, they can carry man to the heights of holy perceptions. Misused and perverted, as they were in Egypt, they can lead man to his destruction.

Because of this relationship between sight and the Plague of Locusts, much of the imagery used to describe the Plague concerns sight:

"And they shall cover the eye (*ain*) of the earth, that one shall not be able to see the earth . . . For they covered the eye of the whole earth, so that the land was darkened."

In truth, it is difficult to understand the significance of not being able to see the earth and the darkening of the land. In fact, because of this difficulty, the *Klei Yakar*, a classic commentary upon the Bible, deviates from Rashi's translation that "whoever wished to, could not see the land" and interprets the passage as follows: " 'the locusts could not see the land because of their own multitude.' Just as the blind eat voraciously and are not easily satisfied because sight is a prominent factor in satisfaction, so the locusts ate without pause or satiation because they could not see how much they were devouring."

The difficulties in this interpretation are self-apparent. Yet, once we appreciate the importance of the sense of sight to the Plague of Locusts, it becomes clear why the Torah stresses that the land could not be seen. As their land was literally consumed before their eyes, and the Egyptians could no longer see their land, some Egyptians began to realize the folly of their misuse of the precious sense of sight. For the first time in the progression of the plagues, the servants of Pharaoh, in other words, the common people, spoke up:

"And Pharaoh's servants said unto him: 'How long shall this man be a snare unto us? let the men go . . . Do you not yet know that Egypt is destroyed?"

THE COMMON PEOPLE SPEAK

The "wise men" and magicians had been discredited long ago. Now, Pharaoh at least listened to the common people and began to "bargain" with Moses:

"And he said unto them: 'Go, serve the Lord your God; but who are they that shall go?"

At this point, Pharaoh was still convinced that even if there *were* spiritual imperatives, they were only for the old, the frail in body or those few who were inclined to the spiritual life. It was not until the next Plague that Pharaoh recognized that the religious life can and *must* involve all ages and members of society. Yet even then, as we shall see, Pharaoh failed to grasp exactly what the spiritual life involved. It was the combination of *Barad* and *Arbeh*, the lessons of the two most important of man's senses, that was beginning to bring Pharaoh to *his* senses, but for him it was too late. He had forfeited his Free Will with the Plague of Boils and he would have to endure two more Plagues so that the Children of Israel could complete their prerequisite course for the Great Event at Mount Sinai.

Chapter XII

DARKNESS: THE LESSON OF SOCIETY

WHO WAS THE SERPENT?

The Plague of Darkness corresponds to the sin of the Serpent in enticing mankind to disobey their Creator. Who or what exactly is the Serpent in the story of the Fall? Let us take a look at the ideal world the Creator had planned for man if he had not transgressed, and the place of the serpent in that world. Unlike many of the non-traditional commentators and critics of the Bible, the Torah recognizes in the Serpent the ultimate representative of the animal world. In the original plan of creation, the Serpent, not the lion, was the king of beasts:

"We thus find it with the primeval Serpent which set its eyes on that which was not proper for it; what it sought was not granted to it and what it possessed was taken from it. The Holy One, blessed is He, said, 'I declared: let it be king over every animal and beast; but now, cursed are you above all cattle and above every beast of the field. I declared: let it walk with an erect posture; but now it shall go upon its belly. I declared: let its food be the same as that of man; but it shall eat dust.' " (Sotah 9b)

SERVANT OF MAN

The Serpent was to use its position as the acme of the animal world and king of the beasts for a specific purpose:

"Rabbi Shimon ben Menassia said: Woe for the loss of a great servant. For had not the serpent been cursed, every Israelite would have had two valuable serpents, sending one to the north and one to the south to bring him precious gems, costly stones and pearls. Moreover, one would have fastened a thong under its tail, with which it would bring forth earth for his garden and wasteland." (Sanhedrin 59b)

Thus, the Talmud paints for us a lavish picture of the ideal human situation in which man would have had at his disposal the wisdom, services and devotion of the king of the animal world, and through him, the entire animal kingdom. The hierarchy of nature was originally set up for just such a purpose, so that man, the zenith of creation, could be served by the rest of the universe, while he dedicated his time and energies to the work of his Creator. Until he sinned, it was this very serpent that was the closest to man, the only one who would dare to converse with him and who represented the rest of creation to him. Yet, it was this very proximity and similarity to man that caused both of their downfalls:

THE SERPENT'S ENVY

"Rabbi Judah be Tema said: Adam reclined in the Garden of Eden while the ministering angels roasted flesh and strained wine for him. Thereupon, the Serpent looked in, saw his glory and became envious of him." (Sanhedrin 59b)

No other animal, beast or fowl was even *capable* of jealousy toward Adam. Envy is based on some factor in common. The pauper envies the rich man for he knows the value of money. The aborigine or cannibal does not envy the multi-millionaire for he has no concept of what money is. So it was with the Serpent and Adam. Having been granted the status of king of the beasts, striding erect and partaking of the same food as Adam, the forces of jealousy began to churn within him. Rashi, in his commentary upon the Torah, brings the Serpent's particular type of envy into even sharper focus:

"*And the Serpent was subtle*: How does this passage belong here? . . . only to teach you for what reason the Serpent pounced upon them, for he saw them engaging in intercourse in the open, and conceived a passion *for them*." Although some texts of Rashi read *for her*, the correct text has been established as reading *for them*. The proper understanding would therefore be, not that the Serpent lusted for Eve, but that he had a consuming desire *to be like Adam and Eve*. In fact, as his passion *to be human* literally consumed him, he decided that he was *better* than them:

"*And the Serpent was subtle*: What was his subtlety? Said he: I know that the Holy One, blessed is He, has decreed, 'in the day that you eat thereof you shall surely die;' I shall go and trick them the man and his wife and they shall eat and be punished, and I will inherit the earth for myself." (Yerushalmi Kiddushin 4:1)

From these words of the Talmud, we may readily see that the Serpent craftily planned mankind's downfall, using also the Creator's laws for the world in his machinations. The Wily Serpent relied on the fact that the Lord would punish Adam and Eve with death and that he, being the "next best thing" would inherit the world. We begin to see that the Serpent becomes identical with what is called in other places the "Evil Inclination" —the *Yetzer Harah.* Wherein lay Eve's error in succumbing to the Serpent's wiles? Surely, this was the first case of enticement and how was Eve to know what was being done to her? The answer lies in the mandate which was given to Adam and shared with Eve upon her creation:

THE UNIQUENESS OF MAN

"And God blessed them; and God said unto them: 'Be fruitful and multiply and replenish the earth and subdue it; *and have dominion* over the fish of the sea, and over the fowl of the air, *and over every living thing* that creeps upon the earth."

Adam and Eve were taught by God to recognize and take to heart the fact they were unique. While respecting the integrity of nature, they had to realize that they had been created to be above all the creatures of creation, for they had been granted the holy soul, "a component of the Divine." This uniqueness carried with it certain imperatives which Eve ignored to her own and mankind's peril. Being created specifically to be elevated to the heights of the sublime, Adam and Eve had to maintain their distance from the beasts of the field. Eve failed to realize that even the most majestic of the beasts *belonged to another world.* When God had breathed the holy soul into Adam, He had forever separated mankind from the animal kingdom. Eve's fatal error therefore lay, not so much in being convinced by the Serpent's casuistry, *but in listening to him at all.* Were Eve's self-respect absolutely intact, her attitude should have been, "how dare you speak with me at all, let alone concerning matters that do not concern the subjects of *your* kingdom, the animal world." But unfortunately, Eve did listen, and the result is still with us.

OUR FRIENDS WE CHOOSE

In essence, the story of the Serpent, his own sin and his enticement of Eve, can be summarized as the Transgression of Society. Man forms many relationships in a lifetime. Some he inherits, such as those with his family etc., and he can either treasure them or tolerate them, depending on his own situation. But there are other relationships which are up to each individual to establish for himself. Among the most important of these is the society in which he chooses to move. His friends, both close and casual, acquaintances, colleagues etc., all leave their mark upon the individual's character This is the lesson to be learned from Eve's brief but fateful association with the serpent.

THE PUNISHMENT FITS THE CRIME

We may now appreciate the justice of the sentences upon the Serpent and Eve for their transgressions:

"And the Lord God said unto the Serpent: 'Because you have done this, cursed are you from among all cattle, and from among all beasts of the field; . . . And I will make enmity between you and the woman, and between your seed and her seed."

"Unto the woman He said: 'I will greatly multiply your pain and your travail; in pain shall you bring forth children; and your desire shall be to your husband, and he shall rule over you.' "

The Serpent, who had enjoyed the trust and companionship of mankind, was suddenly transformed into the most odious member of the animal kingdom. Not only was he no longer the king of beasts, but even the dignity of the beasts was taken away from him. Eve, who had compromised the independence of the human species and had not recognized its unique nature, had her own independence taken away. Not only would she not be allowed to be independent of her husband, but she would not want to.

THREE DAYS OF REFLECTION

What was the essence of the Plague of Darkness?

"They saw not one another, neither rose any from his place

for three days." As Rabbi S. R. Hirsh puts it, "every man was completely cut off from his fellow man." The message was clear: when the light once again returns and you set about to re-establish your relationships, do so in the bright light, not in the darkness of the Serpent. Had the Children of Israel been learning their lessons as they emerged from the influence of Egypt? The Torah gives us a strong hint:

"But all the children of Israel had light *b'moshvotam*—in their dwelling places."

FAMILIES GRADUATE TOGETHER

The Hebrew word *moshav* is used here to denote dwelling place, rather than the more common *batei.* The reason may perhaps be that *moshav* includes the connotations of society, neighborhood, etc. The light that the Jews enjoyed in their dwelling places was granted them because they had in fact learned the essential lesson of proper and improper society. While the Egyptians could not even communicate with each other, the Jewish people would soon culminate their "graduation" from the Egyptian university with a meal that had to be eaten *in a group*. The *Korban Pesah,* the Paschal Sacrifice, symbolized and apotheosized all that the Jews had learned and absorbed during the Ten Plagues. The central aspect of the *Korban Pesah* was therefore its familial nature:

"In the tenth day of this month they shall take to them every man a lamb, according to their fathers' houses, a lamb for a household; and if the household be too little for a lamb, then shall he and his neighbor next unto his house take one according to the number of the souls."

Thus, as they enjoyed the Paschal Offering and reclined at the first Seder, the Jewish people came to realize the value of proper society and strong family ties. The dissolution of the Egyptian family, the haphazardness of Egyptian living arrangements, as we studied them earlier, no longer appealed to the Jewish Nation and indeed all they could feel was a deep loathing and perhaps pity for the Egyptian people.

THE SERPENT AND PHARAOH

On a more symbolic level, we may recall that Pharaoh is himself referred to as the serpent of the Nile:

"Behold, I am against you, Pharaoh King of Egypt, the great dragon that lies in the middle of his rivers, that has said: my river is my own and I have made it for myself."

Just as the Serpent in the Garden of Eden was the wisest of the beasts of the field, so was Pharaoh the personification of all Egyptian wisdom. Pharaoh, like the Primeval Serpent, used his knowledge and wisdom to try to mislead humanity rather than helping them to tread the proper path. Just as the Serpent wished to inherit the earth so did Pharaoh wish to divorce the Jewish people from their destiny and cause them to permanently sink into the immoral Egyptian abyss. As we have established earlier, the Egyptian people did not have to make a conscious effort to influence the Jewish people toward iniquity; it was their very essence to radiate contagious immorality and contamination. But Pharaoh planned and calculated every step in his plot to morally annihilate the Jewish people; and so he is equated with the ancient Serpent.

The Plague of Darkness when Pharaoh lost his powers as surrogate of the ancient Serpent, brought Pharaoh to the concession that the Jews could

"go, serve the Lord; only let your flocks and your herds stay behind; let your little ones also go with you."

TOO PROFOUND FOR PHARAOH

Pharaoh had come a long way since the Plague of Blood. He was now ready to concede not only that there was a need for spiritual work in the world, but that the imperatives of education required the young to participate fully. Yet, Pharaoh still could not grasp the need for including the physical aspects of life in spiritual service. If you are going to give up the Egyptian voluptuous life, Pharaoh argued, you must also give up all contact with the physical world. Leave your flocks and your wealth behind, for you surely will not need them any longer. To this, Moses answered:

"You must also give into our hand sacrifices and burnt-offerings that we may sacrifice unto the Lord our God."

Moses tells Pharaoh, the representative of all wisdom and knowledge: not only will we use our own resources for the service of the Lord, but *you too* will furnish us with media for our Divine vocation. Moses' answer is an eternal one and applies to each generation and its scientific and technological achievements. The Torah does not want us to ignore progress; on the contrary, every new invention and discovery can have one purpose only—to be utilized in the service of the Almighty. Thus, the tape-recorder and record player, camera and movie projector, airplane and automobile etc., can all be used to enhance and improve our worship and labor for the Lord. Yes, Pharaoh, you too shall provide us with weapons in the battle to bring light to the world through the darkness which *you* helped to create. But first, you and your people will have to learn the final lesson, which will remain in men's hearts forever.

Chapter XIII

THE PLAGUE OF THE FIRST BORN

THE PARADOX

The Plague of the Firstborn is unique among the Plagues in that it does not correspond to any one era in history or to any single sin of mankind. Its meaning is universal and its theme is timeless. This concept can be brought out most clearly by examining a few of the extraordinary aspects of the Plague.

Upon reading about the Plague of the Firstborn in the Torah, we discover an interesting paradox. On the one hand, the Lord promised that He alone would carry out the killing of the firstborn:

"I will pass through the land of Egypt that night; I, Myself, and not an angel. And I will smite every firstborn in the land of Egypt; I, myself and no Seraph; And against all the gods of Egypt will I execute judgments; I, myself, and not a messenger."

Yet, on the other hand, in the first nine Plagues, the Jews did not have to do anything in order to be spared from the Plague. Only in the Plague of the Firstborn did the Jews have to make an effort—the placing of the blood on the doorpost—in order not to be smitten along with the Egyptians. It seems strange that the only Plague which the Lord brought about by Himself—without Moses or Aaron—should require the active participation of the Children of Israel.

This paradox can best be explicated by recognizing the dual character of the Plague of the Firstborn. We know that within each of the ternates—*D'tzakh*, *Adash*, *B'ahav*—the first two Plagues in each group were preceeded by a warning. In each case, the warning came immediately before the Plague. Only the Plague of the Firstborn was mentioned by Moses to Pharaoh at the very beginning of their dealings:

"And you shall say unto Pharaoh: Thus says the Lord: Israel is my son, My firstborn. And I have said unto you: Let My son go, that he may serve Me; and you have refused to let him go. Behold, I will slay your son, your firstborn" (Exodus 4:22-23).

The Plague of the Firstborn clearly served two functions: It was the last of the Ten Plagues, and thus Pharaoh was warned of its coming as with the last Plague of each group. Yet, in addition, the Plague of the Firstborn was the symbolic and actual culmination of the long spiritual battle between Egypt and Israel. Egypt, personified by Pharaoh, the "great Serpent," represents the Primeval Serpent's forces of materialism and iniquity in the world. Just as the Serpent was once the king of beasts but had his kingdom taken away because of his actions, so was Egypt once the proudest culture, the most highly developed civilization and the potential leader of the world morally as well as technologically.

THE PARTNERSHIP

The Firstborn is the symbol of all that is best and most cherished. In Hebrew, the word *b'khor*—firstborn— is related to *bikurim*—first fruits—and represents the first, both chronologically and qualitatively. The Egyptians took the best that *they* had, their great wisdom, creativity and industry, and dedicated it to the forces of evil. Israel, the antithesis of Egypt, dedicates all Firsts to the Lord. Thus, it was absolutely imperative that the Jews *participate* in the Plague of the Firstborn and not be mere bystanders, rescued by the benevolent Creator. The Plague of the Firstborn marked the beginning of the great *partnership* between Israel and the Lord. As we learned earlier, from Sinai and onward there would be no more "handouts." Israel would have to "earn its keep" in the world by performing the work of the Lord. That sacred work began with the Plague of the Firstborn.

CONCLUSION

Now that we have traveled back through time, to the most depraved of ages and to the most exciting, to the wisest and the most foolish, we must certainly feel a powerful sense of *deja vu.* Yet, it is not only that we have experienced all this before, but that we are living through all this once again, NOW. Can we claim that our own governmental systems are not plagued by the *tzfardeah* which repudiates common sense and prudence? Have we not all been witness to great and knowledgeable people leading those who are less educated astray? Have we not, only recently, awoken to the large-scale pollution of our rivers, streets, and worst of all, our minds? Have we not abused our bodies until they have rebelled and plagued us with all manner of hitherto unknown ills?

It must be manifest that not only have we not yet emerged from the Iron Furnace, but that it is being rapidly reheated to an intolerable degree. The Midrash tells us that before the Messiah comes, Edom, the symbol of all iniquity before the Messianic Age, will be plagued by the same calamities as those which afflicted Egypt. Surely, that is an indication that the Ten Plagues were not merely one-time punishments for an ancient enemy. They remain eternal lessons from which we can pattern our lives and structure our goals, so that they match and coincide with the Creator's Divine plan for the world. If we do so, we shall undoubtedly reap the same reward as did the Frogs in Egypt, who sanctified the name of the Lord, and were granted the most precious gift of all, that of life. Let us all extricate ourselves from the Iron Furnace and prepare to "delight in the Lord . . . and ride upon the high places of the earth" (Isaiah 58: 14).

SOURCES

Page

Kiddush: Rambam, Hilkhot Shabbat, 29:1.

15 *Maimonides*: Hilkhot Hametz Umatzah 7:1 and K'riat Shma 1:3.
Recent Rabbinical Authorities: See *Shaagat Aryeh*, Responsa numbers 8 and 10 and *Sheelat David* by Rabbi David Friedman of Karlin, #1 and *Or Sameah* by Rabbi Meir Simhah Hacohen of Dvinsk to Hilkhot Kriat Shmah 1:3. See also *L'or Hahalakha*, pp. 221-222.
Psalm 111:4: For a more in depth treatment of this thought, see *Pahad Yitzhak* by Rabbi Isaac Hutner, volume on Passover, I, 21:6.

18 Cyrus H. Gordon, *The Ancient Near East* (New York: W. W. Norton & Co. Inc., 1965), p. 57.
D. Winton Thomas, ed., *Documents From Old Testament Times*, J. M. Plumley, "The Teaching of Amenemope," (New York: Harper & Row, 1961), 172.
Shulhan Arukh: Orah Haim 473:5 and Mishneh Berurah, *Ibid*, 42.

19 *The Midrash**: Yalkut Shimoni 1:163.

20 *Patriotism*: See Avot 3:2, Gittin 10b, Nedarim 28a etc.
Jacob afraid of being worshipped as an idol: Rashi to Genesis 47:29.
Holy Zohar: Zohar Hadash, Parshat Yitro.

25 *The Midrash*: Bamidbar Rabah 22:3.
Another Midrash: Shemoth Rabbah 14:3.

26 *The Midrash*: Mekhilta to Exodus 13:18.

29 *Torah not a history book*: See, for instance Rabbi Joseph Rosen's (known as the Rogatshover Gaon) assertion that "there are no stories in the Torah, only living eternal guidelines through Halakha" (Mihktevei Torah, p. 99.).

30 *The Midrash*: Bamidbar Rabbah 21:4.

31 *The Torah*: See, for instance, Berakhot 57b, Sotah 49a, Pesikta Rabbati, Exodus 17:14, and Deuteronomy 25:17 etc.

32 *Traditional Jewish clothing in Egypt*: Shir Hashirim Rabbah, 4:24. See also Yalkut Shimoni 1:664 and 768 and Rabbi Reuven Margolies, *Ollot*, p. 59.

33 *Vast Divine plan*: Hullin 7b.
Prototype of all future exiles: This thought is found in many traditional books on Jewish theology dealing with the Exodus. See, especially, *Gevurot Hashem* by the Maharal of Prague, *Or Zarua Latzadik* by Rabbi Tzadok Hacohen of Lublin and *Pahad Yitzhak* by Rabbi Isaac Hutner, 71:4.

34 *The future Redemption*: Yalkut Shimoni 1:199.
Rabbi Dessler: *Mikhtav Maeliyahu* 2:20.

37 *Adam and Eve*: See Bava Batra 58a, Hagigah 12a, Sanhedrin 38b, Eruvin 18a, Rashi to Genesis 4:25 etc.

38 *Torah given to men, not angels*: See Deuteronomy 30:12.

39 *Manna gave each person the taste he desired*: Shemoth Rabbah 25:3.

*Unless otherwise noted, the "Midrash" refers to the Midrash Rabbah on the quoted Biblical passage. Biblical commentaries for which references are not given will also be found upon the quoted text.

40 *Manna fell according to personal merit*: Yumah 75a.
41 *Manna remembered on the Sabbath*: Sabbath 117a, Tosaphoth Pesahim 100b, s.v. "Sheein;" see also Eruvin 43b, Betza 16a and Sanhedrin 38b.
43 *As shown in the Talmud*: Megillah 14b, Sotah 11b, Yevamot 65a, Sabbath 118b, etc.
44 *Jabal prayed to his idol in secret*: Yalkut Meam Loez 1:190.
The strains of the universal music: Yalkut Shimoni 2:22.
Jubal prayed to his idol openly: Yalkut Meam Loez 1:190.
46 *Mankind labored with great hardship*: Rashi to Genesis 5:29.
51 *Flood explained away by Dor Hahaflaga*: Rashi to Genesis 11:1.
Rabbi Dessler: *Mikhtav Maeliyahu* 1:184, 3:43 and 3:96.
53 *Nimrod*: Pesahim 54b, Hullin 89a, Avodah Zarah 53b, Eruvin 53a.
Maimonides: Hilkhot Avodah Zarah, Chapter I.
55 *Isaac*: Beraishit Rabbah 56:11, 94:5 and Berakhot 62b.
The Midrash: Tanhumah, Parshat Kedoshim.
56 *All of Jacob's children righteous*: Yalkut Shimoni 1:833.
57 *The Midrash*: Shemoth Rabbah 3:16.
58 *Rabbi Joseph Rosen*: *Tzaphnat Paaneah* to Exodus 4:10.
59 *Aaron as the personification of goodness*: Tanhumah, Parshat Hukat and Yalkut Hadash.
70 *"Blood" brought the Jews great wealth*: Shemoth Rabbah 9:10.
79 *Egypt immoral in sexual matters*: Kiddushin 49b.
81-82 *The Halakhot of who may go to war*: Yevamot 65b, Sotah 44b, Kiddushin 2b etc.
83 *Educational system of Egypt*: Shemoth Rabbah 9:4.
The black magic of the hartumim: Yalkut Shimoni 1:168, 235; Menahot 85a, Shekalim 3:3; Sotah 22a, Zevahim 62b, Targum Yonatan to Numbers 22:22 etc.
86 *Commentators and Responsa Writers*: Rabbi Yonatan Eyebeshutz, *Yaarot Devash* 1:22; *Shvut Yaakov* 2:106; *Torah Shebaal Peh* Volume 14, pp. 46-65
97 *Our sages*: Sanhedrin 108a.
99 *The Midrash*: Tanhumah Kedoshim.
102 *Adam*: Rashi to Genesis 2:19.
110 *The names of God*: See Shavuot 35a, and *Or Hahaim* to Exodus 3:14.
113 *Code of Jewish Laws*: Hoshen Mishpat 420:31 and Shulhan Arukh Harav 5:4 (*Nizkei Haguf*).
If he confesses to a crime: See Radbaz to Rambam, Hilkhot Sanhedrin 18:6.
114 *Istakel B'oreita*: Zohar, Terumah.
derekh haemtzait: Rambam, Hilkhot Deiot.
115 *Jewish Law*: Sanhedrin 36b.
116 *Midrash*: Shemoth Rabbah 11:5.
120 *Good health*: See, for instance, Yevamot 64b, Tamid 27b etc.
128 *The Talmud*: Sanhedrin 38b.
142 *The Midrash*: Tanhumah Yashan, Parshat Bo.

INDEX